Digital Audio Broadcasting

Principles and Applications

Digital Audio Broadcasting

Principles and Applications

Edited by

Wolfgang Hoeg
Berlin, Germany
(formerly head of division at Deutsche Telekom Berkom)

Thomas Lauterbach
University of Applied Sciences, Nürnberg, Germany

JOHN WILEY & SONS, LTD
Chichester · New York · Weinheim · Brisbane · Singapore · Toronto

Copyright © 2001 by John Wiley & Sons, Ltd
                    Baffins Lane, Chichester,
                    West Sussex, PO19 1UD, England

                    National 01243 779777
                    International (+44) 1243 779777
e-mail (for orders and customer service enquiries): cs-books@wiley.co.uk

Visit our Home Page on http://www.wiley.co.uk or http://www.wiley.com

Other Wiley Editorial Offices

John Wiley & Sons, Inc., 605 Third Avenue,
New York, NY 10158-0012, USA

Wiley-VCH Verlag GmbH
Pappelallee 3, D-69469 Weinheim, Germany

John Wiley Australia Ltd, 33 Park Road, Milton,
Queensland 4064, Australia

John Wiley & Sons (Canada) Ltd, 22 Worcester Road
Rexdale, Ontario, M9W 1L1, Canada

John Wiley & Sons (Asia) Pte Ltd, 2 Clementi Loop #02-01,
Jin Xing Distripark, Singapore 129809

Library of Congress Cataloging-in-Publication Data
Digital audio broadcasting : principles and applications / edited by Wolfgang Hoeg,
Thomas Lauterbach
       p. cm.
    Includes bibliographical references and index.
    ISBN 0 471 85894 3
       1. Digital audio broadcasting I. Hoeg, Wolfgang II. Lauterbach, Thomas

    TK6562. D54 D54 2001
    621.384—dc21   00-069340

British Library Cataloguing in Publication Data

A catalogue record for this book is available from the British Library

ISBN 0 471 85894 3

Produced from Word files supplied by the authors
Printed and bound in Great Britain by Antony Rowe Ltd., Chippenham, Wiltshire
This book is printed on acid-free paper responsibly manufactured from sustainable forestry,
in which at least two trees are planted for each one used for paper production.

Contents

Foreword

Many of you who will read this book will do so in the context of an engineering or technical background. And that's a good thing because the more engineers and technical operations people understand this brilliant technology the quicker it will be adopted in countries that have yet to make their digital radio transition strategies. But it is equally important that the content providers and programmers understand what this platform can provide them in terms of service and commercial opportunities as we begin the 21st century...the Digital Era.

I'm a programmer by background and it was the programming opportunities afforded by Digital Audio Broadcasting (DAB, Eureka 147) that led me to believe it would be instrumental in maintaining Radio's relevance in a world of digital choices. Interference free and CD like sound, particularly attractive in the urban mobile environment, are only the beginnings of the DAB advantages. It is spectrum efficient, flexible in terms of bit rate to program content and cheaper to operate than analog systems.

The added bonus to DAB is that the content provider can add value for the listener with program and non-program associated data. Everything from simple song titles and performer to weather, news and traffic flashes are capable of being displayed today. Soon we will have integrated IT systems for the car or home capable of sophisticated navigation, GPS and local DAB traffic and travel information capable of interaction through GSM or phone return. This technology will also facilitate the introduction of full multi media services, again for the mobile or stationary market. What an opportunity for traditional radio broadcasters to bring new services and commercial ventures to their audiences.

DAB is a relatively mature technology, over ten years old. But in the first 7 or 8 years most services were experimental and most receivers were prototype. But in the last couple of years real operation services have begun. In Europe, the UK has an aggressive plan for national and regional multiplexes that will cover about 85% of the population within two years. Today in London 35 digital services are on the air, half of which are only available through DAB.

Germany likewise has an aggressive plan that will see 85% of their population covered within three years. Spain, Portugal, the Nordic countries, and the Benelux countries all are on the air or have aggressive plans to be operational in the next year or two. France is debating how to move from experimental services to full operational activity. And receivers for the car, home and soon hand–held mobile devices are and will be increasingly available in the marketplace. As this book is published, over 16 receiver models and 3 DAB PC cards are available and more are coming to market. Prices were high and still are at premium levels but they are steadily falling to affordable prices.

DAB is important in the context of Europe. Since the technology was developed in Europe, the rest of the world is looking to Europe for successful transition strategies and DAB adoption. But from the beginning countries like Canada (currently operational in 4 major cities) and Australia (planning operational services 2001/02) found the technology innovative and right for their markets. Singapore began operations two years ago, Taiwan China and Hong Kong all have experimental multiplexes. Brunei, Malaysia, Korea and India are all considering experimental services in the near future. Many other countries are assessing the technology and making DAB plans based on their cultural and market needs.

I think those of us who have been part of DAB for almost the last decade were surprised that there wasn't instant takeoff and overnight success for DAB services. We were probably naïve, because this is a replacement technology for analog radio which is ubiquitous around the world and the magnitude of change required is enormous. We began to understand this in the mid nineties and formed the World DAB Forum late in 1996 to facilitate the transition on a Global basis. It is a remarkable group of industry stakeholders including broadcasters, network providers, regulators, consumer electronic manufacturers, electronic business and internet related companies and research institutes.

They have a common purpose: to seek solutions to spectrum and regulatory issues, service availability, commercial technical applications, receiver diversity and availability, and the best practices for market transition. We share best practices nationally, regionally and globally and our agenda changes to meet the current needs and issues of the day. Our goal is to facilitate an orderly transition from analog to DAB radio, ensuring the future relevance, cultural and economic well being of radio.

It is with this backdrop to the potential of DAB and the current operational commitments to DAB services around the world that I commend this technical review of the DAB system in the hope that more knowledge about this wonderful system will generate more services and more countries making operational commitments.

Michael McEwen,   London, Autumn 2000
President, World DAB Forum

Preface

The new digital radio system DAB (Digital Audio Broadcasting), developed within the Eureka 147 project in close co-operation with the EBU, is a very innovative and universal multimedia broadcast system which is just being introduced and has the potential to replace the existing AM and FM audio broadcast services in many parts of the world in the near future. In several countries in Europe and overseas, broadcasting organisations, network providers and receiver manufacturers are already implementing digital broadcasting services using the DAB system.

DAB is very different from conventional analogue broadcasting systems. Most of the system components such as perceptual audio coding (MPEG-1/2), OFDM channel coding and modulation, the provision of a multiplex of several services and data transmission protocols (MOT) are new concepts typical of digital broadcasting. Even experts in analogue transmission systems will feel less familiar with these new elements of broadcasting technology. Therefore, the aim of this book is to inform the expert reader about the basic concepts of the DAB system

Besides introducing the basics, the focus of the book is on the practical implications of service provision and the new infrastructure required in broadcasting houses, for multiplex and network management, and for coverage planning. Also some elements of up-to-date receiver concepts are described.

The level of standardisation of the DAB system is rather advanced and the relevant recent international standards and related documents are introduced and referred to for easy access for the reader seeking technical details. An extended bibliography is also provided.

The book is designed as a well-structured technical guide by a team of expert authors closely involved in the development and standardisation of DAB. This ensures competent presentation and interpretation of the facts based on the latest state of the art. The book is primarily aimed at professional users such as developers and manufacturers of professional devices for distribution networks or consumer receivers, planning engineers and operational staff with broadcasters, network providers, service and content providers. For other technically minded people who wish to become acquainted with the concepts of digital broadcasting, the book will serve as a comprehensive introduction to the field since it contains all the information needed for further study.

The book may also serve for academic or educational use because it is based on the latest versions of the relevant international standards and publications, as well as actual experience with pilot applications and first implementation of services.

The editors wish to take this opportunity to express their thanks to all the contributors for the enjoyable co-operation and their excellent work, which most of them had to complete in addition to their demanding jobs. Many thanks also to Mrs Helga Schön who was kind enough to design a portion of the drawings for the book.

The editors also wish to thank the publishing team at John Wiley & Sons for their interest, understanding and patience during the writing and production period.

May this book help to introduce the DAB system world-wide.

The Editors   Berlin/Nürnberg, Autumn 2000

Copyright Acknowledgement

List of contributors

Editors

WOLFGANG HOEG, Dipl.-Ing., AES Fellow; Berlin, Germany,
graduated from the University of Technology, Dresden, in 1959 and joined the research centre RFZ of Deutsche Post (Deutsche Telekom since 1991). He later became head of the "Audiosystems" division with Deutsche Telekom Berkom, Berlin. He has worked in various fields of audio engineering, such as multichannel audio, DAB and other new broadcasting technologies. He was a member of standardisation bodies of OIRT, ITU-R and EBU, contributed to Eureka 147/DAB Working Groups and chaired Task group DRC.

THOMAS LAUTERBACH, Prof. Dr. rer. nat., Nürnberg, Germany,
received his Diplomphysiker degree and Ph.D. from Erlangen University. In 1992 he joined Robert Bosch, where he got involved with the development DAB. In 1997 he became head of a multimedia systems development department. He was with several Eureka 147/DAB Working Groups, the German DAB platform and EuroDAB/WorldDAB and contributed to ETSI. He also co-ordinated the MEMO (ACTS) project. Since 1997 he is with the Georg-Simon-Ohm-Fachhochschule Nürnberg – University of Applied Sciences as a Professor of Physics. He is currently involved in the Digital Radio Mondiale (DRM) project.

Contributors

STEPHEN BAILY, M.A. (Cantab); London, United Kingdom,
joined the British Broadcasting Corporation in 1985. For the last six years he has worked for BBC Research and Development on various aspects of digital broadcasting, with a particular focus on transmission aspects. He has designed experimental and operational transmission equipment for DAB, and has also worked in the areas of satellite delivery, spectrum planning and design of broadcast systems. He has been a contributor to several Eureka 147/DAB Working Groups, Task Forces and the Executive Committee.

MICHAEL BOLLE, Dr.-Ing., Dresden, Germany,
received his Dr.-Ing. degree with honors from Ruhr-Universität Bochum. Since 1992 he has been with Robert Bosch GmbH working in the field of DAB receiver development. Since 1997 he has been Head of the Department "Advanced Development Multimedia Systems" in which the Bosch DAB VLSI project took place. Michael Bolle left Bosch in 1999 and is now co-founder and executive VP of engineering of Systemonic AG in Dresden. He holds a number of patents in the field of DAB receiver technology.

THOMAS BOLTZE, Dr.-Ing., AES, SMPTE and IEEE, Eindhoven, The Netherlands,
obtained his Ph.-D. in electronics engineering from the University of Technology in Hamburg, Germany. In 1995 he joined the Advanced Systems and Applications Laboratory of Philips Consumer Electronics in Eindhoven. Since, he worked as Senior Systems Designer on audio compression systems and contributed to ITU, AES, Eureka 147/DAB and most notably to the DVB project. He is currently in charge of a joint co-operation with the Research Institute of Television and Electro Acoustics of the MII in Beijing, China.

DETLEF CLAWIN, Dr.-Ing., Pacifica, USA,
received his Dr.-Ing. degree from Ruhr-Universität Bochum. He worked in the fields of high speed circuit design for fiber optics and RFICs for DAB receivers, at Bell Communications Research in Redbank, NJ, USA, and in Germany at Fraunhofer-Gesellschaft IMS in Duisburg, MICRAM Microelectronics in Bochum, and Robert Bosch Corporate Research in Hildesheim, responsible for the development for a single chip DAB tuner IC. Presently, he is a resident engineer at the Stanford Networking Research Center.

BERNHARD FEITEN, Dr.-Ing., AES, Berlin, Germany,
received his doctor degree in electronics at the Technische Universität Berlin in the field of psycho-acoustic and audio bit-rate reduction. He worked as an assistant professor at the Technische Universität Berlin in communication science. Since 1996 he is with Deutsche Telekom. At the research company T-Nova Deutsche Telekom Berkom, he is now head of division „Audiosystems", amongst others responsible for current research on DAB matters. He is member of the ISO/MPEG Audio group.

EGON MEIER-ENGELEN, Dipl.-Ing., Cologne, Germany,
received his Dipl.-Ing. degree in Communications Engineering from the Technische Hochschule Aachen in 1963. Since he has followed a career in research and development on various fields of communications technologies with ITT/SEL, Stuttgart, Exatest Messtechnik GmbH, Leverkusen, and Philips Kommunikations Industrie, Cologne. In 1985 he joined the German Aerospace Center (DLR) in Cologne where he heads a section managing research grants for information technology by the Federal Ministry for Education and Research (BMBF). He directed the Eureka 147 / DAB project from 1986 to 1998.

TORSTEN MLASKO, Dipl.-Ing., Pattensen, Germany
received the Dipl.-Ing. degree in Electrical Engineering from the University of Hanover in 1995. Since 1995, he is with Robert Bosch GmbH, Department Advanced Development Multimedia Systems, Hildesheim, Germany. He is now as a group manager co-ordinating the activities in the field of digital transmission systems, like DAB and DRM, including the development of DAB chip-sets. He was active member of Eureka 147 Working group A and of ISO/MPEG Audio group and is presently active member of WorldDAB and DRM.

HANS-JÖRG NOWOTTNE, Dr.-Ing., Dresden, Germany,
studied Electrical Engineering at University of Technology in Dresden and worked for a long time in R&D of electronic industry focussed on computer-aided design. From 1992 he is with the Fraunhofer-Institute of Integrated Circuits, Dresden. As group manager he deals with prototype development in the field of telecommunications and digital broadcasting and is involved in DAB since 1993. In the Eureka147/DAB project he has been involved in the definition of the Service Transport Interface (STI) and was with Working group B.

ROLAND PLANKENBÜHLER, Dr.-Ing., Nürnberg, Germany,
studied Electrical Engineering at the University of Erlangen. After his Diploma Degree he worked there at"Lab for Technical Electronics". Having obtained his Ph.D. he joined Fraunhofer IIS-A in 1990 and became manager of the "Data Services" group which was involved with DAB. Since 1997 he manages the Department "Terminal Devices which is mainly focussed on definition and implementation of data services and terminals. He was a member of several working groups of the Eureka DAB as well as WorldDAB Module A.

THOMAS SCHIERBAUM, Munich, Germany,
after his technical education, he studied media marketing at the Bayerische Akademie für Werbung und Marketing. He is with the Institut für Rundfunktechnik (IRT), division Informations- und Datendienste, working as a project manager for data service controlling and ancillary data transmission in DAB. He was responsible for infrastructure of the DAB-Program "Bayern Mobil" of the Bayerischer Rundfunk, DAB product management and consulting. He is with several Working groups of the German public broadcasters.

HENRIK SCHULZE, Prof. Dr. rer. nat, Meschede, Germany,
received the masters degree (Dipl.-Phys.) in 1983 and his Ph.D. in 1987 from the University of Göttingen, Germany. From 1987 to 1993 he was with the Robert Bosch GmbH at the Research Institute in Hildesheim, where he worked on the development of modulation and channel coding for the DAB system. Since 1993 he is Professor for Communication Theory at the University of Paderborn, Division Meschede. He is currently involved in the Digital Radio Mondiale (DRM) project.

GERHARD STOLL, Dipl.-Ing., AES Fellow, München, Germany,
studied electrical engineering at the universities of Stuttgart and Munich. In 1984 he joined the Institut für Rundfunktechnik (IRT) and became head of the psycho-acoustics group. He was responsible for the development of the MPEG-Audio Layer II standard. Mr Stoll was/is also a member of different international standardisation groups, such as MPEG, Eureka 147/DAB, DVB and EBU. As a senior engineer at the IRT, he is now in charge of advanced multimedia broadcasting and information services.

WOLFRAM TITZE, Dr., AMIEE, Berlin, Germany
graduated from Friedrich-Alexander University Erlangen and received his PhD in Electronic Engineering from University College London in 1993. He joined Robert Bosch and became later head of the department multimedia systems development working on aspects of DAB. He was member of the Eureka 147/DAB Executive Committee, worked on standardisation of DAB distribution interfaces at ETSI and was vice chairman of the Technical module of WorldDAB. He also worked as Eureka 147 promotion engineer. Dr. Titze is now director of product development of the Cortologic AG.

LOTHAR TÜMPFEL, Dipl.-Ing., Berlin, Germany,
studied Telecommunication Engineering in Leipzig and Information Technology in Dresden. He worked for many years on the field of development of audio studio equipment. Since the 1990s he was with the Deutsche Telekom, now at T-Nova Deutsche Telekom Berkom dealing with research and development for DAB. He was member of several Working groups and of the Steering Committee of Eureka 147/DAB, and contributed to the EBU Specialist Group V4/RSM.

HERMAN VAN VELTHOVEN, Ing., AES; Antwerpen, Belgium,
joined Pioneer Europe in 1973 where he became successively Manager Product Planning General Audio and Manager Engineering Department. In 1996 he changed to Pioneer's R&D Division. When in 1993 Pioneer joined the Eureka 147/DAB project as the first non-European company, he represented his company in several Working groups and Task forces. He now represents Pioneer in WorldDAB Module A. He chaired EACEM PT1.1 and was also an ETSI rapporteur. He is member of CENELEC TC-206.

Abbreviations

A/D	Analogue/Digital
AAC	Advanced Audio Coding
ACI	Adjacent Channel Interference
ACTS	Advanced Communications Technologies and Services
ADC	Analogue-to-Digital Converter
ADR	Astra Digital Radio
ADSL	Asymmetric Digital Subscriber Line
AFC	Automatic Frequency Control
AGC	Automatic Gain Control
AGL	Above Ground Level
AIC	Auxiliary Information Channel
AM	Amplitude Modulation
API	Application Programming Interface
ASCII	American Standard Code for Information Interchange
ASCTy	Audio Service Component Type
Asu	Announcement Support
ATM	Asynchronous Transfer Mode
AWGN	White Gaussian Noise
BAL	Bit Allocation
BER	Bit Error Rate
BST	Band Segmented Transmission
CA	Conditional Access
CAZAC	Constant Amplitude Zero Autocorrelation
CD	Compact Disk
CDMA	Code Division Multiple Access
CENELEC	Comité Européen de Normalisation Electrotechnique
CEPT	Conférence des Administrations Européenes des Postes et Telecommunications
CIF	Common Interleaved Frame
CMOS	Complementary Metal Oxide Semiconductor
COFDM	Coded Orthogonal Frequency Division Multiplex
CRC	Cyclic Redundancy Check
CU	Capacity Unit
CW	Control Word
D/A	Digital/Analogue
DAB	Digital Audio Broadcasting
DAC	Digital-to-Analogue Converter
DAD	Destination Address
DAT	Digital Audio Tape
dBFS	Relative signal level in decibels, related to the digital coding limit 0 dBFS (Full Scale)
dBrel	Relative signal level in decibels, related to a pre-defined reference level
dBu	Absolute signal level in decibels, related to a reference voltage of 0 dBu = 0.775 V
DCA	Digitally Controlled Amplifier
DE	Downstream Entity
DECT	Digital European Cordless Telephone
DGPS	Differential GPS
DIQ	Digital In-phase and Quadrature
DMB	Digital Multimedia Broadcast
DQPSK	Differential Quadrature Phase Shift Keying
DRC	Dynamic Range Control
DRCD	DRC Data
DRM	Digital Radio Mondiale
DSCTy	Data Service Component Type
DSP	Digital Signal Processor

DSR	Digital Satellite Radio
DTB	Desktop Publishing System
DVB(-T)	Digital Video Broadcasting(-Terrestrial)
EACEM	European Association of Consumer Electronics Manufacturers
EBU	European Broadcasting Union
EC	European Community
ECC	Extended Country Code
ECM	Entitlement Checking Message
EEP	Equal Error Protection
EId	Ensemble Identifier
EMC	Electromagnetic Compatibility
EMM	Entitlement Management Message
EN	European Standard (ETSI document type; normative)
EOH	End Of Header
EPP	Ensemble Provider Profile
ERP	Effective Radiated Power
ETI	Ensemble Transport Interface
ETS	European Telecommunications Standard (ETSI document type; normative)
ETSI	European Telecommunications Standards Institute
EWS	Emergency Warning System
FCC	Federal Communications Commission
FEC	Forward Error Correction
FFT	Fast Fourier Transform
FI	Frequency Information
FIB	Fast Information Block
FIC	Fast Information Channel
FIDC	Fast Information Data Channel
FIG	Fast Information Group
FM	Frequency Modulation
F-PAD	Fixed Programme-Associated Data (PAD)
FTP	File Transfer Protocol
GIF	Graphics Interchange Format
GPS	Global Positioning System
GSM	Global System for Mobile
HF	High Frequency
HiFi	High Fidelity
HTML	Hyper Text Markup Language
I/Q	In-phase and Quadrature
IBAC	In-Band-Adjacent-Channel
IBOC	In-Band-On-Channel
Id	Identifier
IF	Intermediate Frequency
IFA	Internationale Funkausstellung (World of Consumer Electronics), Berlin
IFFT	Inverse Fast Fourier Transform
IM	Intermodulation
IMDCT	Inverse Modified Discrete Cosine Transformation
IP	Internet Protocol
IS	International Standard (ISO document type; normative)
ISDB	Integrated Services Digital Broadcasting
ISDN	Integrated Services Digital Network
ISO	International Organisation for Standardization
ITTS	Interactive Text Transmission System
ITU-R	International Telecommunications Union, Radiocommunications Sector
ITU-T	International Telecommunications Union, Telecommunications Standardization Sector

JESSI	Joint European Submicron Silicon Initiative
JPEG	Joint Photographic Experts Group
LAN	Local Area Network
LI	Logical Interface
LSB	Least Significant Bit
LTO	Local Time Offset
M/S	Music/Speech
M4M	Multimedia for Mobiles
MAC	Multiple Accumulate
MCI	Multiplex Configuration Information
MEMO	Multimedia Environment for Mobiles
MFN	Multiple Frequency Network
MJD	Modified Julian Date
MNSC	Multiplex Network Signalling Channel
MOT	Multimedia Object Transfer
M-PAD	Motion PAD
MPEG	Moving Pictures Experts Group
MSB	Most Significant Bit
MSC	Main Service Channel
MST	Main Stream Data
MTBF	Mean Time Between Failures
MUSICAM	Masking pattern-adapted Universal Sub-band Integrated Coding and Multiplexing
MW	Medium Wave
NAB	National Association of Broadcasters (USA)
NASC	Network-Adapted Signalling Channel
NICAM	Near Instantaneously Companded Audio Multiplex
NRSC	National Radio Systems Committee
ODA	Open Data Application
OFDM	Orthogonal Frequency Division Multiplex
OSI	Open Systems Interconnection
PAD	Programme-Associated Data
PAL	Phase Alternating Line
PCM	Pulse Code Modulation
PDH	Plesiochronous Digital Hierarchy
PI	Programme Identifier
PL	Protection Level
PLL	Phase-Locked Loop
PMC	Production technology Management Committee (EBU)
PNum	Programme Number
PPM	Peak Programme Level Meter
pps	pulse per second
PRBS	Pseudo-Random Binary Sequence
PRC	Prime Rate Channel
PSK	Phase Shift Keying
PSTN	Public Switched Telephone Network
PTy	Programme Type
QAM	Quadrature Amplitude Modulation
QPSK	Quaternary Phase Shift Keying
RAM	Random Access Memory
RBDS	Radio Broadcast Data System
RCPC	Rate Compatible Punctured Convolutional Codes
RDI	Radio Data Interface/Receiver Data Interface
RDS	Radio Data System
RF	Radio Frequency

RISC	Reduced Instruction Set Computer
RMS	Root Mean Square
RS	Reed Solomon
SAD	Start Address
SAW	Surface Acoustic Wave
SC	Synchronisation Channel
SC Lang	Service Component Language
SCF-CRC	Scale Factor CRC
SCFSI	Scale Factor Select Information
SCT	Service Component Trigger
SDH	Synchronous Digital Hierarchy
SES	Société Européenne des Satellites
SFN	Single Frequency Network
SI	Service Information
SId	Service Identifier
SMPTE	Society of Motion Pictures & Television Engineers
SNR	Signal-to-Noise Ratio
SPP	Service Provider Profile
SSTC	Single Stream Characterisation
STC-C(TA)	STI Control Transport Adapted
STI	Service Transport Interface
STI(PI,X)	STI Physical Interface
STI-C	STI Control
STI-C(LI)	STI Control Logical Interface
STI-D	STI Data
STI-D(LI)	STI Data Logical Interface
TA	Traffic Announcement
TCP	Transfer Control Protocol
TDC	Transparent Data Channel
TDM	Time-Domain Multiplex
TFPR	Time Frequency Phase Reference
TII	Transmitter Identification Information
TM	Transmission Mode
TMC	Traffic Message Channel
TP	Traffic Programme
TPEG	Transport Protocol Experts Group
TR	Technical Report (ETSI document type; informative)
TS	Technical Specification (ETSI document type; normative)
TTI	Traffic and Traveller Information
TV	Television
UE	Upstream Entity
UEP	Unequal Error Protection
UHF	Ultra High Frequency
UMTS	Universal Mobile Telecommunication System
UTC	Co-ordinated Universal Time
VCA	Voltage Controlled Amplifier
VCO	Voltage Controlled Oscillator
VHF	Very High Frequency
VLSI	Very Large Scale Integration
VU	Volume Unit
WARC	World Administrative Radio Conference
WCDMA	Wide-Band Code Division Multiple Access
WFA	Wave Digital Filter
X-PAD	Extended Programme Associated Data (PAD)

1

Introduction

WOLFGANG HOEG, THOMAS LAUTERBACH,

EGON MEIER-ENGELEN and HENRIK SCHULZE

1.1　　General

The new digital radio system DAB (Digital Audio Broadcasting) is a very innovative and universal multimedia broadcast system which will replace the existing AM and FM audio broadcast services in many parts of the world in the future. It was developed in the 1990s by the Eureka 147/DAB project. DAB is very well suited for mobile reception and provides very high robustness against multipath reception. It allows use of single frequency networks (SFNs) for high frequency efficiency.

Besides high-quality digital audio services (mono, two-channel or multichannel stereophonic), DAB is able to transmit programme-associated data and a multiplex of other data services (e.g. travel and traffic information, still and moving pictures, etc.). A dynamic multiplex management on the network side opens up possibilities for flexible programming.

In several countries in Europe and overseas broadcast organisations, network providers and receiver manufacturers are going to implement digital broadcasting services using the DAB system in pilot projects and public services.

DAB works very differently from conventional broadcasting systems. Most of the system components such as perceptual audio coding, channel coding and modulation, multiplex management or data transmission protocols are new solutions and typically not so familiar to the expert in existing analogue or digital broadcast systems.

The level of standardisation of the DAB system is rather advanced and the various recent international standards and related documents are introduced and referred to for easy access for the reader seeking technical details.

Digital Audio Broadcasting: Principles and Applications, edited by W. Hoeg and T. Lauterbach
©2001 John Wiley & Sons, Ltd.

1.2   Radio in the Digital Age

Radio broadcasting is one of the most widespread electronic mass media comprising hundreds of programme providers, thousands of HF transmitters and billions of receivers world-wide. Since the beginning of broadcasting in the early 1920s, the market has been widely covered by the AM and FM audio broadcasting services.

Today we live in a world of digital communication systems and services. Essential parts of the production processes in radio houses were changed to digital ones in recent times, beginning with the change from conventional analogue audio tape to digital recording on tape, hard disk or CD, via digital signal processing in mixing desks and digital transmission links in distribution processes. The first steps in the introduction of digital broadcasting services were taken by the systems NICAM 728 (Near Instantaneously Companded Audio Multiplex, developed by the BBC for stereo television sound in the VHF/UHF bands), DSR (Digital Satellite Radio), or ADR (Astra Digital Radio, see section 1.6.2), but they were not suited to replace the existing conventional services completely, especially for mobile reception.

Normally, it takes a period of a human generation (or at least a period in the life of a receiver type generation, i.e. approximately 10 years) to replace an existing broadcasting system by a new one. Therefore, strong reasons and very convincing advantages are required to justify the introduction of a new broadcasting system.

1.3   Benefits of the Eureka 147 DAB System

However, there will always be some problems, or additional effort will be needed when replacing an existing technology by a new one, such as

- lack of transmission frequencies
- costs for development and investment
- looking for providers for new non-conventional services (e.g. data services)
- solving the chicken and egg problem (who will be first – the service provider or the receiver manufacturer?).

Nevertheless, the new Eureka 147 DAB system provides a wealth of advantages over conventional audio broadcast systems such as analogue VHF/FM or AM radio, and also partly over other existing digital broadcast systems such as DSR, ADR, etc. The following list will only highlight some key advantages as an overview; many more details will be explained in the corresponding sections of the book.

Quality of service
DAB uses all the possibilities of modern digital communication technologies and can thus provide a much higher level of quality of service, such as

- *Superior sound quality:* DAB users can enjoy pure undistorted sound close to CD quality. New features such as Dynamic Range Control (DRC) or Music/Speech Control can be used individually by customers to match the audio quality to their needs.
- *Usability:* Rather than searching wavebands, users can select all available stations or preferred formats from a simple text menu.

- *Perfect reception conditions:* With just a simple, non-directional whip antenna, DAB eliminates interference and the problem of multipath while in a car. It covers wide geographical areas with an even, uninterrupted signal. Once full services are up and running, a driver will be able to cross an entire country and stay tuned to the same station with no signal fade and without altering frequency.

Wide range of value-added services
DAB is quite unique in that both music and data services can be received using the same receiver. One receiver does it all, such as

- *Typical audio broadcasting (main service):* Music, drama, news, information, etc., can be received in monophonic or stereophonic form as is well known from conventional radio programmes; there is also the potential to transmit multichannel (5.1 format) audio programmes as well.
- *Programme-associated data (PAD):* DAB broadcast receivers can display text information in far greater detail than RDS, such as programme background facts, a menu of future broadcasts and complementary advertising information. Receivers attached to a small screen will display visual information such as weather maps or CD cover images.
- *Information services:* Services from sources other than the broadcasting station are included within the same channel for the user to access at will. These include news headlines, detailed weather information or even the latest stock marked prices.
- *Targeted music or data services:* Because digital technology can carry a massive amount of information, specific user groups can be targeted with great accuracy because each receiver can be addressable.
- *Still or moving pictures:* Data can also appear as still or moving photographic pictures, accompanied by an audio service or as separate information.

Universal system layout
The DAB system has a fairly universal and well-standardised system layout which allows applications for all known transmission media and receiving situations.

- *Standardisation:* The level of international standardisation of all basic principles and transmission tools for the new DAB system is very high (more than 40 international standards cover all necessary details).
- *Unique system design:* DAB services will be available manly on terrestrial but also suited for cable and satellite networks, and the same receiver could be used to provide radio programmes and/or data services for national, regional, local and international coverage.
- *Wide choice of receivers:* It is possible to access DAB services on a wide range of receiving equipment including fixed (stationary), mobile and portable radio receivers, optionally completed with displays or screens, and even personal computers.

Flexibility of multiplex configuration
DAB services are transmitted in a flexible multiplex configuration, which can be easily changed instantaneously to the actual needs of the content providers.

- *Multiplex configuration:* The arrangement of services in a DAB multiplex may be changed instantaneously to match the needs of the providers of programmes or data services, without interrupting ongoing services.
- *Bit rate flexibility:* The programme provider can choose an appropriate bit rate for a certain audio programme according to its quality, for instance less than 100 kbit/s for a pure speech programme, 128 kbit/s for monophonic or 256 kbit/s for stereophonic music; also half sampling frequency can be used for lower quality services. So the available bit rate can be split optimally between different services of a DAB ensemble.

Transmission efficiency
Compared to conventional broadcast systems much less economic effort in investment and operation is needed for a DAB transmission system.

- *Lower transmission costs for broadcasters:* DAB allows broadcasters to provide a wide range of programme material simultaneously on the same frequency. This not only makes room for a vastly increased number of programmes to increase user choice, but also has important broadcast cost-cutting implications.
- *Lower transmission costs for transmitter network providers:* For digital transmission a DAB transmitter needs only a fraction of the electrical energy compared to a conventional AM or FM transmitter.
- *Frequency efficiency:* DAB transmitter networks can be designed as Single Frequency Network (SFNs), which saves a lot of transmission frequencies and thus transmission capacity on air.

These advantages of DAB (and there are more if we look further into the details) justify the introduction of DAB into the media world in order to replace the existing conventional radio systems step by step over a longer period.

1.4   History of the Origins of DAB

1.4.1   Steps of Development

In the early 1980s the first digital sound broadcasting systems providing CD like audio quality were developed for satellite delivery. These systems made use of the broadcasting bands in the 10 to 12 GHz region, employed very little sound data compression and were not aimed at mobile reception. Thus, it was not possible to serve a great majority of listeners, such as those travelling in cars. Also, another feature of the well-established FM radio could not be provided by satellite delivery, namely "local services". Consequently terrestrial digital sound broadcasting was considered as an essential delivery method for reaching all listeners.

At first investigations were initiated by radio research institutes looking into the feasibility of applying digital modulation schemes in the FM bands. However, the straightforward use of pulse code modulation (PCM) in the upper portions of the FM band generated intolerable interference in most existing FM receivers and was spectrally very inefficient. Mobile reception was never tried and would not have succeeded. A much more sophisticated approach was definitely necessary.

In Germany the Federal Ministry for Research and Technology (BMFT, now BMBF) launched a research initiative to assess the feasibility of terrestrial digital sound broadcasting comprising more effective methods of sound data compression and efficient use of the radio spectrum. A study completed in 1984 indicated that promising results could be expected from highly demanding research activities. As a new digital sound broadcasting system could only be implemented successfully by wide international agreement, BMFT set the task for its Project Management Agency at DLR (German Aerospace Centre) to form a European consortium of industry, broadcasters, network providers, research centres and academia for the development of a new digital audio broadcasting system. Towards the end of 1986 a consortium of 19 organisations from France, Germany, The Netherlands and the United Kingdom had signed a co-operation agreement and applied for notification as a Eureka project. At the meeting in December 1986 of the High Level Representatives of the Eureka partner states in Stockholm the project, now called "Digital Audio Broadcasting, DAB", was notified as the Eureka147 project. National research grants were awarded to that project in France, Germany and The Netherlands. However, owing to granting procedures official work on the project could not start before the beginning of 1988 and was supposed to run for four years.

Credit must also be given to the European Broadcasting Union (EBU), which had launched work on the satellite delivery of digital sound broadcasting to mobiles in the frequency range between 1 and 3 GHz, by awarding a research contract to the Centre Commun d'Etudes de Télédiffusion et Télécommunications (CCETT) in Rennes, France, prior to the forming of the DAB consortium. As the CCETT also joined the DAB project, the work already begun for the EBU became part of the DAB activities and the EBU a close ally and active promoter for DAB. Later, this proved very important and helpful in relations with the International Telecommunications Union (ITU-R) and the standardisation process with the European Telecommunications Standards Institute (ETSI).

From the beginning the goals set for the project were very demanding and difficult to achieve. Perfect mobile reception was the overall aim. In detail the list of requirements to be met included the following items:

- audio quality comparable to that of the CD
- unimpaired mobile reception in a car, even at high speeds
- efficient frequency spectrum utilisation
- transmission capacity for ancillary data
- low transmitting power
- terrestrial, cable and satellite delivery options
- easy-to-operate receivers
- European or better world-wide standardisation.

The first system approach considered at least 16 stereo programmes of CD audio quality plus ancillary data to be transmitted in the 7 MHz bandwidth of a television channel. This definitely cannot be achieved by simply transmitting the combined net bit rates of 16 CD-like programme channels, which are around 1,4 Mbit/s each, over the TV channel. So a high degree of audio data compression without any perceptible loss of audio quality was mandatory. Data rates below 200 kbit/s per stereo channel had to be achieved.

Unimpaired mobile reception was also required to overcome the adverse effects of multipath signal propagation with the associated frequency selective fading.

Audio data compression and the transmission method became the central efforts of the research project. Both tasks were addressed in a broad and comprehensive manner. For audio coding four different approaches were investigated: two sub-band coding systems competed with two transform coding systems. Similarly, for the transmission method four different schemes were proposed:

- one narrow-band system
- one single carrier spread-spectrum system
- one multicarrier OFDM system
- and one frequency-hopping system.

All approaches were developed to an extent where – either through experimental evidence or at least by thorough simulation – a fair and valid comparison of the performance of the proposed solutions became possible. The period of selection of and decision for the best suited audio coding system and the most appropriate transmission scheme was a crucial moment in the history of the Eureka 147 consortium. For audio coding the greatest part of the selection process happened external to the consortium. All four coding schemes previously had been within the activities of the ISO/IEC Moving Pictures Experts Group (MPEG), which worked on standardisation for data compressed video and audio coding. There the solutions offered by Eureka 147 competed against 10 other entries from other countries, world-wide. The MPEG Audio Group set up a very elaborate and qualified audio quality assessment campaign that was strongly supported by Swedish Radio, the British Broadcasting Corporation and the Communications Research Centre, Canada, among others. The very thorough subjective audio quality tests revealed that the audio coding systems submitted by Eureka 147 showed superior performance and consequently were standardised by ISO/IEC as MPEG Audio Layers I, II and III. Within the Eureka 147 consortium, after long consideration Layer II, also known as MUSICAM, was selected for the DAB specification.

The process of choosing the most appropriate transmission method took place within the Eureka 147 consortium alone. In simulations performed according to rules worked out by the members the four approaches were put to the test. This showed that the broadband solutions performed better than the narrow-band proposal. Among the broadband versions the spread-spectrum approach had a slight advantage over the OFDM approach, while the frequency-hopping solution was considered too demanding with respect to network organisation. However, the OFDM system was the only one that was already available in hardware with field-test experience – in the form of the coded Orthogonal Frequency Division Multiplex (COFDM) system, while the spread-spectrum proposal by then was not developed as hardware at all and estimated to be very complex. So, the choice fell on COFDM, which has since proven to be an excellent performer.

A further important decision had to be made relating to the bandwidth of the DAB system. From a network and service area planning point of view as well as obtainable frequency spectrum perspectives, an ensemble of 16 programmes on one transmitter with the 7 MHz bandwidth of a TV channel proved to be much too inflexible, although in experiments it had provided very good performance in a multipath environment. Therefore, a considerable but reasonable reduction in transmission bandwidth was necessary. In Canada experiments with the COFDM system revealed that substantial performance degradation begins around 1.3 MHz and lower. So, a reasonable bandwidth for a DAB channel or "DAB block" was defined as 1.5 MHz. This allows several possibilities, as follows.

A 7 MHz TV channel can be divided into four DAB blocks, each carrying ensembles of five to seven programmes. With four blocks fitting into 7 MHz service area planning is possible with only one TV channel, without having adjacent areas using the same DAB block. Furthermore, 1.5 MHz bandwidth is sufficient to transport one MPEG coded audio/video bit stream.

After the above-mentioned important decisions had been made the members of the consortium all turned their efforts from their individual approaches to the commonly defined system architecture and with rapid progress developed the details of the complete basic DAB specification to be submitted to the international standardisation bodies. By that time, another European research project, the JESSI flagship project AE-14, was eagerly awaiting the DAB specification to begin the development of chip-sets for DAB. Also, the standardisation bodies like ETSI were well aware and waiting for the submission of the specification since Eureka 147 members had been very active in testing and presenting the results of their research and development on many important occasions, together with or organised by the EBU.

The first official presentation of DAB took place at the World Administrative Radio Conference 1988 (WARC'88) in Geneva for the delegates of this conference. As the issue of a frequency allocation for digital sound broadcasting in the L-band around 1.5 GHz was up for decision, a demonstration simulating satellite reception was presented. A transmitter on Mont Salève radiated the DAB signal in the UHF TV band, giving in downtown Geneva an angle of signal incidence similar to that from a geostationary satellite. Mobile reception was perfect and the delegates to the conference were highly impressed. In consequence the conference assigned spectrum in the L-band for satellite sound broadcasting with terrestrial augmentation permitted.

One year later DAB was presented at the ITU-COM exhibition in Geneva. In 1990 tests with mobile demonstrations were run by Canadian broadcasters in Toronto, Ottawa, Montreal and Vancouver. DAB demonstrations and exhibits were shown at all International Radio Shows (IFA) in Berlin and the UK Radio Festivals in Birmingham since 1991. Four International DAB Symposia have been held up to now: 1992 in Montreux, 1994 in Toronto, 1996 again in Montreux and 1999 in Singapore. In 1994 a mobile DAB demonstration was presented at the Arab States Broadcasting Union Conference in Tunis. From all these activities world-wide recognition and appreciation was gained for DAB.

The efforts and results of finding acceptance for DAB in the United States deserve an extra paragraph. As early as 1990 – by invitation of the National Association of Broadcasters (NAB) – the consortium presented DAB at low key at the NAB Convention in Atlanta, Georgia. This led to a very elaborate mobile demonstration and exhibition at the next NAB Convention in 1991 in Las Vegas, Nevada. Several of the NAB officials by that time were very interested in reaching a co-operation and licence agreement with the Eureka 147 consortium. However, strong opposition against that system – requiring new spectrum and the bundling of several programmes onto one transmitter – also arose. The US radio industry – that is, the broadcasters – feared new competition from the licensing of new spectrum. They preferred the idea of a system approach named "In-Band-On-Channel (IBOC)", where the digital presentation of their analogue programmes is transmitted together and within the spectrum mask of their licensed FM channel (see also section 1.6). This of course would avoid the need of new licensing for digital broadcasting and thus keep new competition away. However appealing and spectrum efficient this concept may be, the realisation might prove to be a very formidable technical task. No feasible development was available at that time, but fast development of an IBOC system was promised. Eureka 147 DAB performed flawlessly at Las Vegas, but those opposing the system claimed that the topography around Las Vegas was much too favourable to put DAB to a real test. So it was requested that DAB should next come to the 1991 NAB Radio Show in San Francisco to be tested in a very difficult propagation environment. One main transmitter and one gap filler were set up and mobile reception in downtown San Francisco was impressively demonstrated. An announced demonstration of IBOC broadcasting did not take place as the equipment was not ready. In spite of the good results the opposition to DAB gained momentum and the NAB officially announced its preference for an IBOC system. This was not in line with the intentions of the Consumer Electronics Manufacturers Association (CEMA) of the Electronics Industry Association (EIA) which came to an agreement with the NAB to run very strictly monitored laboratory tests in Cleveland and field trials in San Francisco comparing the Eureka 147 DAB with several US IBOC and IBAC (In-Band-Adjacent Channel) systems. The results were to be presented to the Federal Communications Commission (FCC) for rule-making relating to digital sound broadcasting. While Eureka 147 soon had equipment ready for the tests, the US proponents for a long time could not provide equipment and delayed the tests until 1995. In the laboratory tests DAB outperformed all competing systems by far. Claiming to have been unfairly treated in the evaluation of the laboratory tests, all but one of the US proponents of terrestrial systems withdrew from the field tests in San Francisco. For Eureka 147 the Canadian partner Digital Radio Research Inc. (DRRI) installed on contract a single frequency network of one main transmitter and two gap fillers to provide coverage for the area designated for mobile testing. Again DAB provided excellent performance as was documented in the final reports of the CEMA. Nevertheless, the United States still pursued the concept of IBOC although several generations of IBOC equipment and redesigns have only produced marginal performance. Even though numerous Americans now admit that DAB is a superior system they claim that it is not suited for the US broadcasters. The FCC has not made any decision on terrestrial digital audio broadcasting up to now.

Standardisation in Europe and world-wide occurred at a better pace. The first DAB-related standard was achieved for audio coding in 1993, when the International Organisation for Standardisation/International Electrical Commission (ISO/IEC) released the International Standard (IS 11172-3) comprising MPEG/Audio Layers I, II and III (IS 11172). Also in 1993 ETSI adopted the basic DAB standard ETS 300 401, with several additional standards following later. The ITU-R in 1994 issued Recommendations (BS.1114) and (BO.1130) relating to satellite and terrestrial digital audio broadcasting, recommending the use of Eureka 147 DAB mentioned as "Digital System A". Consumer equipment manufacturers have also achieved several standards concerning the basic requirements for DAB receivers issued by the Comité Européen de Normalisation Electrotechnique (CENELEC) (for more detailed information see section 1.5).

Even though the technology, the norms and standards had been developed, the most critical issue was still not resolved : provision of frequency spectrum for DAB. WARC'88 had allocated 40 MHz of spectrum in the L-band to satellite sound broadcasting, allowing also terrestrial augmentation. Through the intense intervention of several national delegations WARC'92 conceded primary terrestrial use of a portion of that allocated spectrum, which for several countries is the only frequency resource available. The L-band

– very well suited for satellite delivery of DAB – on the other hand becomes very costly for terrestrial network implementation. VHF and UHF are much more cost efficient than the terrestrial L-band. There was no hope of acquiring any additional spectrum below 1 GHz outside of the bands already allocated to broadcasting. So, in 1995 the Conference Européenne des Administrations des Postes et des Télécommunications (CEPT) convened a spectrum planning conference for terrestrial DAB in Wiesbaden, Germany, that worked out an allotment plan for DAB in VHF Band III (mostly former TV channel 12) and in the L-band from 1452 to 1467 MHz, allowing for all CEPT member states two coverages of DAB, one in the VHF range, the other in the L-band. This decision made possible the installation of experimental DAB pilot services in many countries of Europe and the beginning of regular services starting in 1997 with Sweden and the United Kingdom. More and more countries – also outside of Europe – are following suit. It is recognised that presently the available spectrum is not sufficient to move all existing FM and future programmes to DAB.

1.4.2 Organisations and Platforms

A few of the organisations and bodies were or still are busy promoting and supporting the development and introduction of the DAB system world-wide:

Eureka 147 consortium
As mentioned above the Eureka 147 consortium was the driving force in the development of DAB. It formed a Programme Board, dealing with strategic planning, contractual and legal affairs, membership and promotion, a Steering Committee, planning the tasks of the working groups and making all technical decisions, and four working groups of varying task assignments. DLR in Cologne, Germany, was chosen to act as the managing agency and as the project office for Eureka affairs.

While the original consortium did not accept new members at the beginning, this policy was changed when the basic specification of DAB (EN 300401) was ready to be released to the public for standardisation in 1992. The Eureka 147 Programme Board established rules for the entry of new partners into the consortium, requiring, for example, an entrance fee of DM 150,000 from industrial companies while entry for broadcasters and network operators was free of charge. The money earned from the entrance fees was mainly reserved for system promotion and to a smaller extent for organisational expenses. In 1993 the first new members were admitted and soon the consortium grew to 54 members from 14 countries.

The research work of the original partners had led to a number of basic patents for DAB, individually held by several members. Consequently the consortium came to the conclusion to offer licences as a package to all necessary intellectual property rights (IPR) through authorised agents, one for receiver matters and another for transmitter and measuring equipment.

In 1997 the idea of merging Eureka 147 with the promoting organisation WorldDAB (see below) was discussed with the result that a gradual merger was adopted and the final merger completed by the beginning of the year 2000. The activities of the EU-147 Project Office were transferred to the WorldDAB Project Office by the end of 1998. With the complete merger at the end of 1999 Eureka 147 ceased to exist. The members of the consortium can now co-operate in WorldDAB Module A, where technical issues of DAB are handled.

National DAB platforms

The first national DAB platform *DAB Plattform e.V.* was initiated by the German Ministry for Education and Research in 1991. Founded as a national platform it soon accepted members from Austria and Switzerland and consequently dropped the word "national" from its name. It soon reached a membership of 52 organisations. The main objective of this platform was the promotion of the DAB system in German-speaking countries. The organisation of and activities for DAB presentations at the IFA events in Berlin were highlights in the promotion programme of the German platform. The members decided to end the existence of DAB Plattform e.V. by autumn 1998 after considering it to have achieved its objectives.

Likewise, national platforms were established in several other European countries. France launched the *Club DAB France*, The Netherlands the *Dutch DAB Foundation* and the United Kingdom the *UK DAB Forum*. Promotional activities for the public and information to governmental agencies were again the objectives. At the time of writing these organisations continue to exist.

EuroDAB/WorldDAB Forum

Strongly stimulated and supported by the EBU, an organised co-operation of national platforms and other interested bodies lead to the founding of the *EuroDAB Forum* in 1995. A EuroDAB Project Office was set up at EBU Headquarters in Geneva. Membership rose quickly, bringing together interested organisations not only from Europe but from many parts of the world. Accordingly the General Assembly of EuroDAB decided to change its name to *WorldDAB Forum* in 1997. WorldDAB has around 100 members from 25 countries world-wide. Promotion of DAB is again the main objective. Formerly EuroDAB and now WorldDAB issue a quarterly *DAB Newsletter*. The Forum has organised all but the first International DAB Symposia mentioned above. Technical, legal and regulatory as well as promotional issues are dealt with in several modules of the Forum. Since January 2000 the Eureka 147 consortium has fully merged with the WorldDAB Forum. The WorldDAB Project Office moved from Geneva to London in 1998 and has also acted for Eureka 147 since 1999. An extensive web-site of WorldDAB and DAB information is maintained at this office (www.worlddab.org).

1.4.3   Milestones of Introduction

At the time of writing (mid 2000) DAB remains – after more than 10 years' development and standardisation – still in the phase of early introduction. However, in some regions programme and service providers have already started regular DAB services. Although DAB was primarily developed in the Eureka 147 project as a European broadcast system, it can be shown already in this phase that most of the industrial developed regions world-wide (i.e. Asia, Australia, and parts of America such as Canada and South America) are now interested in introducing this highly sophisticated universal broadcast system, too (see Appendix 2). The world-wide penetration of DAB is mainly supported by the WorldDAB Forum (see above). The actual status of introduction can be found at the web-site also given above.

Only in the important market places of United States and Japan were being developed proprietary digital radio systems, see section 1.6.

1.5   International Standardisation

The new DAB system shows a very high level of standardisation in its principles and applications, which is rather unusual for a new broadcasting or multimedia system. (There are more than 40 corresponding international standards and related documents.)

After the initial development by some broadcasters and related research institutes, the first international standards were passed by the ITU-R (International Telecommunications Union, Radiocommunications Sector). In the period following the system has been developed within the Eureka 147 project, supported by a wide representation of telecommunication network providers, broadcasters, receiver manufacturers and related research institutes, in close co-operation with the EBU.

In the following some basic standards of the DAB system are listed out. A more complete listing of the corresponding standards and related documents can be found in the Bibliography, which is referred to within various parts of the text.

Basic requirements and system standards

The basic ITU-R Recommendation (BS.774) shortly specifies as *"Digital system A"* the main requirements to the new broadcasting system DAB. Other ITU-R Recommendations (BS.789) and (BO.1130) regulate conditions needed for additional frequency ranges for emission.

Several ITU World Administrative Conferences (WARC'79, WARC'85, WARC'88 WARC'92) dealt with digital radio and allocated frequency bands for satellite and terrestrial digital sound broadcasting. More details were decided in (CEPT, 1995).

As a result of developments within the Eureka 147 project, the *Main DAB Standard* or *DAB Specification* (EN 300401) was approved by ETSI European Telecommunications Standards Institute, which defines the characteristics of the DAB transmission signal, including audio coding, data services, signal and service multiplexing, channel coding and modulation. (The document was formerly often called the *ETS;* now it is a European Standard EN.).

Audio coding

DAB represents one of the most important applications of the generic ISO/MPEG-1 Layer II audio coding scheme (IS 11172). The use of this ISO/IEC coding standard is recommended by the ITU-R in (BS.1115), and certainly in the DAB Specification (EN 300401). DAB is also designed to transmit MPEG-2 Layer II audio (IS 13818), for instance for lower quality half-sample-rate transmission or multichannel audio programmes, see also (ES 201755). Other standards specify procedures and test bit-streams for DAB audio conformance testing (TS 101757), or audio interfaces for transmission within the studio region (IEC 60958), (IEC 61937).

Data services

Owing to the special needs for data services in DAB a new standard for Multimedia Object Transfer (MOT) was created defining the specific transport encoding of data types not specified in (EN 300401) and ensuring interoperability between different data services and application types or equipment of different providers or manufacturers. Additional guidelines are given in the "MOT Rules of operation" (TS 101497), "Broadcast Web-site application" (TS 101498) or "Slide show application" (TS 101499). An important bridge between the well-known radio data service RDS for FM (EN 50067) and the data services in DAB is provided by the European Norm (EN 301700) which defines service referencing from FM-RDS and the use of RDS-ODA (open data applications). A transparent data channel for DAB transmission is described in (TS 101759).

Network and transmission standards
Based on the DAB main standard (EN 300401) additional standards are given to define the DAB multiplex signal formats for distribution (networking) and transmission. This is the so-called Service Transport Interface (STI) (EN 300797) for contribution networks between service providers and broadcast studios, the Ensemble Transport Interface (ETI) (EN 300799), and the Digital baseband In-phase and Quadrature (DIQ) interface (EN 300798) for DAB channel coding using OFDM modulation.

To provide interactive services, transport mechanisms for IP Datagram Tunnelling (ES 201735), Network Independent Protocols (ES 201736) and Interaction Channel Through GSM/PSTN/ISDN/DECT (ES 201737) are defined.

Receiver requirements
Based on the DAB main standard (EN 300401) additional standards are given to define the implementation of DAB receivers. (EN 50248) describes DAB receiver characteristics for consumer equipment for terrestrial, cable and satellite reception. EMC parameters for receivers are identified in EACEM TR-004. (EN 50255) specifies the Radio Data Interface (RDI) between DAB receivers and peripheral data equipment. A special command set to control receivers is described in (EN 50320). (TR 101758) lists general field strength considerations for a DAB system.

Guidelines for implementation and operation
Based on the main standards, more detailed guidelines and rules for implementation and operation are compiled in (TR 101296) as a main guideline document for service providers and manufacturers. A guide to standards, guidelines and bibliography is given in the ETSI Technical Report (TR 101495). A broadcaster's introduction to the implementation of some key DAB system features (BPN 007) is provided by the EBU.

1.6   Relations to Other Digital Broadcasting Systems

In addition to the European DAB system which is mainly covered by this book, several other digital sound broadcasting services exist which have been or are being developed and (partly) introduced.

These systems differ in many aspects (specifications, service complexity, parameters) from the DAB concept. Some are focused more strongly on a single application or service (e.g. stationary reception) or provide lower audio quality levels (such as WorldSpace or DRM). Except for ISDB-T, which uses technologies very similar to DAB, all other systems are not expected to be able to comply with the audio quality and quality of service in mobile reception provided by DAB. Nevertheless the basic concepts and limitations of some of these systems will be introduced briefly.

Also, there are a few cable-based digital radio services, which are mainly derived from existing satellite or terrestrial radio systems (for instance, ADR, DVB, etc.) so it may not be necessary to describe them separately. In general, those systems use QAM schemes in order to achieve high data rates in the limited bandwidth of the cable. This is possible because of the high signal-to-noise ratio available. However, these services have only local importance, depending of the extension of the broadband cable distribution network used.

1.6.1   Satellite-based Digital Radio Systems

Although the DAB system can also be used for satellite transmissions, a number of different proprietary systems have been designed by companies providing direct broadcasting satellite services. Of course, many radio programmes are transmitted via telecommunication satellites to feed relay transmitters or cable head stations. These systems are beyond the scope of this section.

Generally, the system layout in the direct broadcasting systems focuses either on stationary reception with highly directive antennas (e.g. "dishes") in the 11 GHz range (e.g. ADR) or on portable and possibly mobile reception in the L-band (1.5 GHz) and S-band (2.3/2.5/2.6 GHz) allocations (e.g. WorldSpace, XM and Sirius).

Historically, the first system for the direct satellite broadcasting of digital radio services was Digital Satellite Radio (DSR). In contrast to later systems, no sound compression was used but only a slight reduction in data rate as compared to CD by using a scale factor. This service has recently been closed down because of the end of the lifetime of the satellite used (Kopernikus). Therefore, it is not described here. More details can be found in (Schambeck, 1987).

The basic building blocks of modern satellite systems are audio coding, some kind of multiplexing, and modulation. While there are things in common with DAB with respect to the first two aspects, these systems do not use OFDM modulation because of the non-linear travelling wave tube amplifiers on satellites and the low spectral efficiency. Instead, often relatively simple PSK schemes are used, which are considered sufficient because, when using directional receiving antennas, multipath propagation does not occur in satellite channels.

Satellite systems face severe problems when the service is to be extended to mobile reception. Because only line-of-sight operation is possible owing to the limited power available in satellite transmissions, at higher latitudes, where geostationary satellites have low angles of elevation, either terrestrial retransmission in cities and mountainous regions is necessary, or the service has to be provided by several satellites at different azimuths in parallel to fill shaded areas. Another approach is to use satellites in highly inclined elliptical orbits which always show high elevation angles. However, this requires several satellites in the same orbit to provide continuous service and difficult switching procedures have to be performed at hand-over from the descending to the ascending satellite.

1.6.1.1   Astra Digital Radio

The Astra Digital Radio (ADR) system was designed by Société Européenne des Satellites (SES), Luxembourg, to provide digital radio on its geostationary direct TV broadcasting satellites. This co-positioned family of satellites labelled "ASTRA" covers Central Europe operating in the 11 GHz range. Brief system overviews are given by (Hofmeir, 1995) or (Kleine, 1995). A detailed description of the system is available at the ASTRA web-site (www.ASTRA).

The system design is compatible with the sub-carrier scheme for analogue sound transmission within TV transponders. It uses MPEG Layer II sound coding (as does DAB) at a fixed bit rate of 192 kbit/s for a stereo signal. Forward error correction is applied using a punctured convolutional code with code rate 3/4. The data are differentially encoded to provide easy synchronisation. QPSK modulation is used for each of up to 12 ADR sub-carriers which are combined with the analogue TV signal and are FM modulated. The baseband bandwidth of the digital signal is 130 kHz which is the same as for an analogue sound sub-carrier. This allows existing analogue radio services to be replaced by the digital ADR service one by one. It is also possible to use a whole transponder for ADR only. In this case, 48 channels can be accommodated in one transponder.

Additional data can be sent together with the audio signal. ADR uses the auxiliary data field of the MPEG audio frame to do this, but the coding is different from the DAB Programme Associated Data (PAD; see Chapter 2). ADR uses 252 bits per audio frame for this purpose, which results in a net data rate of about 6 kbit/s, because a (7,4) block code is used for error correction. This capacity is flexibly split into "RDS data", "Ancillary Data" and "Control Data". RDS data are coded in the format of the RDS (EN 50067) which is used on a digital sub-carrier on FM sound broadcasts and provides service-related information such as labels and Programme Type (see also section 2.5). The reason for using this format on ADR is that several broadcasters use ADR to feed their terrestrial FM networks. Hence these data can be extracted from the ADR signal and fed to the RDS encoder of the FM transmitter. Ancillary data are used by the broadcaster for in-house purposes. The control data field contains a list of all ADR channels to provide easy tuning and allows transmission of parameters for conditional access if applied.

With respect to DAB, ADR provides the same quality of audio, because the same coding mechanism is used. The number of channels available, however, is much larger. While DAB currently only provides six to seven services in each network, ADR can provide 12 digital radio channels on each transponder in addition to a TV ·channel. Therefore, several hundred channels are available and in use by public and private broadcasters from all over Europe. ADR decoders are integrated in high-end satellite TV receivers and are therefore widespread. The most important drawback of ADR is that it can only be received with stationary receivers using a dish antenna. The flexibility of the system with respect to data transmission and future multimedia extensions is considerably lower than that of DAB.

1.6.1.2   WorldSpace

The WorldSpace digital radio system was designed by the US company WorldSpace Inc., to provide digital radio to developing countries on its three geostationary satellites named "CaribStar" covering Central and South America, "AfriStar" covering Africa, Southern Europe and the Near and Middle East, and "AsiaStar" covering India, China, Japan and South East Asia. The system downlink operates in the L-band. The standard is not public. A description is given by (Sachdev, 1997), for example.

The basic building blocks of the systems are "Prime Rate Channels" (PRCs) which each transmit 16 kbit/s of data. Several of these (typically up to eight, resulting in 128 kbit/s) can be combined to provide a stereo sound channel or data channels. For audio coding, MPEG Layer III is used.

Each satellite serves three coverage areas (spots). For each spot, two multiplex signals containing 96 PRCs (i.e. 12 stereo channels) each in a time domain multiplex (TDM) are available. Each of these data streams is QPSK modulated. One multiplex is completely assembled in a ground station and linked up to the satellite from which it is retransmitted without further processing. The other multiplex is processed on board the satellites by assembling individual PRC uplinks to the TDM signal. The advantage of this is that there are no additional costs to transport the signal from the broadcaster to the uplink station. Moreover, a signal can be broadcast in several spots with only one uplink.

Owing to the lower path loss of the L-band signals as compared to 11 GHz, the WorldSpace system is able to provide significant field strength within the coverage area. Therefore reception is possible with portable receivers outdoors and inside buildings using a low-gain patch or helical antenna at a window where line-of-sight conditions apply. Minor attenuation by foliage or trees may be tolerable.

At the time of writing the AfriStar and AsiaStar satellites are operational and several receiver manufacturers are beginning to market receivers.

With respect to DAB, WorldSpace addresses a completely different broadcasting environment. Clearly, in developing countries digital satellite radio will considerably improve the number and quality of broadcast receptions, because in many parts up to now there is only short-wave coverage. Mobile reception is probably not an urgent need and hence the system is not particularly designed for it.

From the point of view of audio quality, by using MPEG Layer III at 128 kbit/s the quality will be comparable to that of DAB at 160–192 kbit/s. At this bit rate, however, the number of channels per spot is restricted to 24. Owing to the large coverage areas this number may not be sufficient to serve the multilingual and multiethnic audience. Hence, the audio bit rate will have to be reduced in order to increase the number of channels available. Nevertheless the sound quality will in many cases still be comparable at least to mono FM, which is still a major improvement on short-wave AM.

Utilising the concept of PRCs the system is very flexible and can be extended to data transmission and multimedia contents. This is important because the lifetime of the satellites is 15 years.

1.6.1.3   Satellite Systems in the United States

In the United States there is no allocation for digital radio in the L-band, but in the S-band. Currently, two companies are in the process of establishing satellite services targeted at mobile reception and intended to cover the whole of the continental United States (i.e. except Alaska and Hawaii). Sirius Satellite Radio intends to use three satellites on inclined orbits to achieve high elevation angles and terrestrial repeaters in metropolitan areas, see (www.Siriusradio). The other company, XM Satellite Radio, will use two high-powered geostationary satellites at 85° and 115° which both cover the entire continental United States and thus provide spatial diversity to overcome shading, see (www.XMRadio). Nevertheless terrestrial retransmitters will also be needed.

At the time of writing, the first satellite of Siriusradio has been launched successfully. The systems will be available in 2001 and then provide approximately 100 audio channels in a quality comparable to CD. The services will be pay-radio based on a monthly subscription fee. Both companies recently agreed to develop a unified standard enabling the same receiver to access both systems.

With respect to sound quality these systems should be comparable to DAB, and with respect to reliability of service these systems should be comparable to WorldSpace; however, because of the measures taken they should show better performance in mobile and portable reception.

1.6.2   Terrestrial Digital Broadcasting Systems

1.6.2.1   Digital Video Broadcasting – Terrestrial (DVB-T)

The DVB system was developed a few years after the DAB system. On the audio and video coding level as well as on the multiplexing level, it is completely based on the MPEG-2 standard. This is in contrast to the DAB standard, where the audio coding is MPEG, but the multiplex control is independent and specially adapted for the requirements. There are three DVB standards that differ very much in the transmission scheme: DVB-S for satellite (EN 300421), DVB-C for cable (EN 300429), and DVB-T for terrestrial broadcasting (EN 300744). Here we consider only the last one. For a detailed description of all three standards, see (Reimers, 1995).

The DVB-T standard has been designed for stationary and portable terrestrial reception with multipath fading. In contrast to DAB, mobile reception was not required. On the coding and modulation level, many methods have been adopted from the DAB system, where they had already been implemented successfully.

Like DAB, DVB-T uses OFDM with a guard interval. There are two transmission modes. The first is called 8K mode (because it uses an 8192-point FFT) and has a symbol length of the same order as in DAB transmission mode I. It is suited for single frequency networks. The second is called 2K mode (because it uses a 2048-point FFT) and has a symbol length of the same order as in DAB transmission mode II. It is not suited for single frequency networks. The guard interval can be chosen to be 25% of the total symbol length, as for DAB, but also shorter guard intervals are possible. The total bandwidth of about 7.6 MHz is suited for terrestrial 8 MHz TV channels. The system parameters can be scaled for 7 MHz TV channels. A major difference from the DAB system is that DVB-T uses coherent modulation. Different signal constellations between QPSK and 64-QAM are possible. At the receiving side, the channel amplitude and phase have to be estimated. For channel estimation, nearly 10% of the total bandwidth is needed for pilot symbols. The coherent modulation scheme with channel estimation by a Wiener filter is more advanced than the differential demodulation scheme of DAB. It is even more robust against fast fading in mobile reception situations (Hoeher, 1991a), (Schulze, 1998), (Schulze, 1999).

The OFDM transmission scheme for DVB-T includes a frequency interleaver that consists of a simple pseudo-random permutation of the sub-carrier indices. It is very similar to the DAB frequency interleaver. In contrast to DAB, the DVB-T system has no time interleaving. Time interleaving only makes sense for mobile reception, which was not a requirement for the design of the system.

The channel coding is based on the same convolutional codes like the DAB system. Code rates between 1/2 and 7/8 are possible. Unequal error protection (UEP) is not possible for the DVB system, since the data and transmission level are completely separated. As a further consequence, in contrast to DAB, the whole multiplex has to be coded with the same code rate. To reach the very low bit error rates that are required for the video codec, an outer Reed–Solomon (RS) code is applied: blocks of 188 bytes of useful data are coded into blocks of 204 bytes. Between the outer RS code and the inner convolutional code, a (relatively) short byte interleaver has been inserted to break up the burst errors that are produced by the Viterbi decoder.

The data rates that can be carried by the DVB-T multiplex vary from about 5 Mbit/s to about 30 Mbit/s, depending on the modulation (QPSK, 16-QAM or 64-QAM), the code rate (1/2, 2/3, 3/4, 5/6 or 7/8) and the guard interval.

Mobile reception of DVB-T, even though not a specific design feature, has been widely discussed and gave rise to many theoretical investigations and practical experiments. Furthermore, there seems to be a controversy between DAB and DVB adherents that concentrates very much on this question. Unfortunately, even though both systems are very close together, the chance has been missed to design a universal common system. From the technical point of view, we will list some comments on the question of mobile reception:

1. Unlike one might expect, the differential modulation of DAB is not more robust than the coherent modulation for DVB even for severely fast-fading mobile reception. The coherent scheme is more robust and spectrally more efficient, see (Schulze, 1999).
2. For mobile reception in a frequency flat fading situation, the frequency interleaving even over a relatively wide band of 8 MHz is not sufficient. As a consequence, the lack of time interleaving may severely degrade the system in some mobile reception situations, see (Schulze, 1998).
3. Synchronisation losses may sometimes occur for mobile receivers. In contrast to DVB-T, this was taken into account in the design of the DAB system: the null symbol allows a rough frame and symbol synchronisation every 24 ms or a maximum of 96 ms. The FIC can then immediately be decoded and provides the information about the multiplex structure. And, in the (typical) stream mode, the data stream does not need separate synchronisation because it is synchronised to the physical frame structure.
4. Bit errors in the scale factors of the MPEG audio data (*birdies*) are very annoying for the human ear. For a subjective impression this is much worse than ,say, block error in a picture. The DAB system uses an extra CRC (that is not part of the MPEG bit-stream) to detect and to conceal these errors. This is not possible for DVB-T.

As a consequence, the DVB-T system allows mobile reception in many typical environments, but it may fail in some situations where more care has been taken in the DAB system.

1.6.2.2   In-Band-On-Channel Systems

In the United States the broadcasting environment is considerably different from the European one. Technical parameters of the stations (AM or FM, power, antenna height) resulting in different coverage areas and quality of audio service appear to be important economic factors that are vital to the broadcasters and which many of them think have to be preserved in the transition to a digital system. Although the DAB system can provide such features to a certain extent by applying different audio data rates and coding profiles, the idea of multiplexing several stations into one transmitting channel does not seem to be very attractive to some US broadcasters.

In the early 1990s, a number of US companies therefore started to work on alternative approaches to digital audio radio (DAR) broadcasting known as IBOC (In-Band-On-Channel) and IBAC (In-Band-Adjacent-Channel). Systems were developed both for AM and FM stations in MW and VHF Band II. The basic principle is to use low-power digital sideband signals within the spectrum mask of the channel allocated to the station.

Later in the 1990s a number of IBOC/IBAC systems and the Eureka 147 DAB system and a satellite-based system (VOA/JPL) were tested in both the laboratory and field in the San Francisco area. The results of these tests were not encouraging for the IBOC/IBAC approaches, see (Culver, 1996), stating explicitly that "of all systems tested, only the Eureka 147/DAB system offers the audio quality and signal robustness performance that listeners would expect from a new DAR service". Since then work on IBOC systems has nevertheless continued. At the time of writing (mid 2000), the US National Radio Systems Committee (NRSC) is starting a new testing procedure, to which the companies USA Digital Radio and Lucent will contribute IBOC system proposals called iDAB and LDR.

However, independent of the individual system details there are two principle problems with the FM IBOC approach which are difficult to overcome.

Owing to the fact that the IBOC signal is limited in bandwidth to the FCC channel spectrum mask, the bandwidth is not sufficient to overcome frequency-selective fading. This means that in portable or mobile reception a fade of the whole signal can occur. In this case there is hardly a chance to reconstruct the signal even if strong error correction codes and "long" interleaving are used (think of stopping at a traffic light). Therefore, from the point of view of coverage, IBOC systems are not expected to be superior to FM, because they will fail in the same locations as FM does.

Every IBOC signal will to a certain extent impair the analogue signal in the same channel. The effect of this is additional noise on the analogue audio signal and this may affect different analogue receiver circuits in different ways. Therefore the broadcaster has no control of what the digital signal will do to all the different analogue receivers being used. To keep this noise sufficiently low, the digital signal level must be kept far below the level of the analogue signal. Although the digital system requires less signal-to-noise ratio than FM, this means that the coverage of IBOC will be very limited.

1.6.2.3   Integrated Services Digital Broadcasting (Japan)

In Japan, the NHK Science and Technical Research Laboratories have proposed a concept called Integrated Services Digital Broadcasting (ISDB). From this approach a system was created which can be configured for terrestrial and satellite broadcasts of radio, TV and multimedia services, see (Kuroda, 1997), (Nakahara, 1996). For the terrestrial system (ISDB-T) the most important aims are rugged reception with portable and mobile receivers, use of Single Frequency Networks (SFNs) to achieve frequency efficient coverage and to have a flexible scheme in order to be future-proof. To achieve this, a variant of OFDM (see section 2.2) was developed which is called Band Segmented Transmission (BST)-OFDM and means that each signal consists of a number of basic OFDM building blocks called segments with a bandwidth of 432 kHz which are spread over the band where frequencies are available. The OFDM parameters of the systems are similar to transmission mode IV of the DAB system because the system is designed for the UHF range.

A single 432 kHz segment is sufficient to broadcast audio programmes and data, but it is also possible to combine three segments for this purpose. For TV signals 13 segments are combined to form a 5.6 MHz wide signal. In all cases the basic modulation and coding modes of the segments are (D)QPSK, 16QAM and 64QAM, and several codes rates between 1/2 and 7/8 are available. In the 5.6 MHz mode the individual segments may be coded and modulated in different ways to achieve hierarchical transmission capacity. For the audio and video coding the MPEG-2 standard will be used.

In 1998 draft standards on digital terrestrial television and radio based on the ISDB-T concept were established, see (STRL, 1999), and field tests are being performed.

From the point of view of relation to DAB, ISDB-T basically uses the same principles and hence provides similar features as do DAB and DVB-T, respectively. By to using the MPEG-2 standard for coding, the audio and video quality will be comparable. Owing to the similarity of the OFDM parameters used, and similar bandwidth (at least when three segments are combined for audio), the coverage properties will also be similar to those of DAB. The advantage of ISDB-T is that the radio and TV systems are based on the same basic building blocks (although the coding and modulation will be different), which allows for reuse of some circuits in receivers, whereas DAB and DVB-T differ in many details. Another advantage of ISDB-T is the availability of a narrow-band mode (i.e. a single segment) which can deliver one or two radio programmes to small regions or communities, although potentially with some reception problems due to flat fading. With this segmentation ISDB-T can also flexibly use the frequencies not occupied by analogue services in the introductory simulcast phase. Such possibilities are not present in DAB.

1.6.3   *Digital Radio in the Broadcasting Bands below 30 MHz*

The frequency range below 30 MHz (short, medium, long wave) offers the possibility to provide a very large coverage area for broadcasting with only one high-power transmitter and very simple receivers with no directional antennas. Using this frequency range is the easiest – and often practically the only possible – way to cover large countries with low technical infrastructure. On the other hand, traditional broadcasting in this frequency range uses the most antiquated and simplest possible transmission scheme: double-sideband AM with a carrier. From one point of view this is a waste of power and bandwidth efficiency in such a valuable frequency range at a time when every year more and more efficient communication systems are being designed. From a practical point of view, this old technique provides only very poor audio quality and thus finds acceptance only in regions where no other service is available.

The obvious need for a more advanced system working in this frequency range gave rise to a consortium called *Digital Radio Mondiale* (DRM) to develop a world-wide standard for such a system. At the time of writing (mid 2000) the work is progressing to complete the specification and to go on with field tests. Although there are still some parameters to be defined, the basic structure of the system seems to be fixed now.

The available bandwidth in this frequency range is typically very small: the usual channel spacing is 5, 9 or 10 kHz. For the new system, it should be possible to bundle two 5 kHz channels to one 10 kHz channel. For such narrow-band channels, much more data reduction is required than for the DAB system. The *Advanced Audio Coding* (AAC) data reduction scheme of MPEG allows an audio quality comparable to FM at a data rate between 20 and 25 kbit/s, see for example (Dietz, 2000). This would be a great improvement compared to conventional AM quality. But bearing in mind that the DAB system typically transmits little more than 1 Mbit/s (e.g. six audio programmes of 192 kbit/s) in a 1.5 MHz bandwidth, even this low data rate is relatively high compared to the extremely small bandwidth of 10 kHz. This means that the DRM transmission scheme must provide more than twice the bandwidth efficiency compared to DAB.

The physical channel depends on the type of wave propagation, which is very special in this frequency range. It may differ from night to day and it may depend on the solar activity. Also interference caused by human noise plays an important role. In situations where only ground-wave propagation is present, there is a simple white Gaussian noise channel that is relatively easy to handle. But typically the wave propagation is dominated by ionospheric scattering (sky wave). This means that the channel is time and frequency variant – just like the mobile DAB channel (see section 2.1). Travel time differences of the wave of the order of 5 ms may occur which cause severe frequency-selective fading. The motion of the ionosphere causes a time variance in the channel with a Doppler spectrum with a typical bandwidth of the order of 1 Hz.

The DRM channel has similar problems in time- and frequency-selective fading as the DAB channel, so the DRM group decided on the same solution: OFDM (see section 2.1). The system parameters are not finally fixed at the time of writing. A possible choice that may be suited for sky wave propagation could be an OFDM symbol duration of 27 ms and a guard interval of 5.3 ms. With this symbol length, the OFDM symbol would consist of approximately 200 sub-carriers. The requirement of a very high-bandwidth efficiency leads to a much higher signal constellation for each sub-carrier than used for DAB. It turned out that 64-QAM is a good choice. Because of the very large time variance of the channel, a large amount of the total bandwidth (about 1/6) will be needed for channel estimation pilots. It turns out that an overall code rate between 1/2 and 2/3 will be needed for the required data rate. It can be shown (Schulze, 1999) that such a coded 64-QAM OFDM system with a good channel estimator is surprisingly robust in a fading channel even with conventional 64-QAM symbol mapping. For DRM it has been decided to use the multilevel coding approach first proposed by (Imai, 1977). This approach is very similar to that described by (Wörz, 1993) for a coded 8-PSK system. With multilevel 64-QAM, an additional gain of up to 1 or 2 dB can be reached compared to conventional 64-QAM. DRM is the first system that implements such a coded modulation scheme, see also (Dietz, 2000). The same convolutional codes as DAB will be used here.

For ground mode propagation, another transmission mode may be defined, because less channel estimation is needed, the guard interval may be shorter, and even weaker codes can be used in a Gaussian channel.

Channel coding must be supported by interleaving to work in a fading channel. Frequency interleaving alone is not sufficient, especially if the echoes are too short or if only a two-path propagation situation occurs. Time interleaving is restricted by the delay that is allowed by the application: after the receiver has been switched on the listener cannot wait too long for a signal. A delay of 2 seconds is regarded as an upper limit. Since the time variance is very slow (1 Hz Doppler spread), the interleaving will not be sufficient in many situations.

1.6.4   Web-casting

A completely different way to provide radio services is the use of the Internet as a universal multimedia platform.

The basic equipment for an Internet radio customer is a PC, complete with a sound adapter, a rather fast modem and Internet access, possibly via ISDN or ASDL.

The availability of improved and scalable audio coding schemes, on the other hand, is creating the opportunity for a better definition of audio quality obtainable via the Web. The Internet's impact on broadcasters has changed from its initial position as a site for posting information related to the station and programme to an additional medium to reach new customers through streaming audio, live or on-demand, still and moving pictures, and new interactive concepts. The importance of Internet audio services, which are using only a small part of the capacity of the net, is already significantly increasing audio information delivered via IP. Several software companies have produced the necessary tools to distribute audio and video via the Internet with the goal to develop the Web as a new mass medium, similar to the traditional audio and video broadcasting services.

With the current streaming techniques and 56 kbit/s modems an audio quality can be achieved surpassing by far that of MW broadcasting with audio bit rates which are reduced by a factor of 50, compared to those used with the CD.

For both, either for live streaming or downloading of audio clips, there are several audio players (software codecs) in use which can provide sufficient audio quality in the range mentioned below, such as

- Microsoft Windows Media 4
- RealNetworks G2
- Q-Design Music Codec 2 together with Quicktime streaming
- MPEG-2/-4 AAC "Advanced Audio Coding"
- MP2, which is MPEG-1/-2 Layer II
- MP3 (close to MPEG-1/-2 Layer III)
- Yamaha SoundVQ.

With this variety of audio players already on the market (mostly for free!) it may be difficult to reach international standardisation in this field. Nevertheless, web-casting will have its own place in the radio world because of the access to hundreds of radio programmes world-wide.

Much more detailed information on web-casting and its audio quality expectations is contained in (BPN 022) and (Stoll, 2000).

2

System Concept

THOMAS LAUTERBACH, HENRIK SCHULZE and
HERMAN VAN VELTHOVEN

2.1   The Physical Channel

Mobile reception without disturbance was the basic requirement for the development of the DAB system.

The special problems of mobile reception are caused by multipath propagation: the electromagnetic wave will be scattered, diffracted, reflected and reaches the antenna in various ways as an incoherent superposition of many signals with different travel times. This leads to an interference pattern that depends on the frequency and the location or – for a mobile receiver - the time.

The mobile receiver moves through an interference pattern that changes within microseconds and that varies over the transmission bandwidth. One says that the mobile radio channel is characterised by time variance and frequency selectivity.

The time variance is determined by the vehicle speed v and the wavelength $\lambda = c/f_0$, where f_0 is the transmission frequency and c the velocity of light. The relevant physical quantity is the maximum Doppler frequency shift:

$$f_{D\max} = \frac{v}{c} f_0 \approx \frac{1}{1080} \frac{f_0}{\text{MHz}} \frac{v}{\text{km/h}} .$$

(2.1)

Table 2.1 shows some practically relevant figures for $f_{D\max}$.

Digital Audio Broadcasting: Principles and Applications, edited by W. Hoeg and T. Lauterbach
©2001 John Wiley & Sons, Ltd.

Table 2.1   Examples for Doppler frequencies

f_{Dmax}	v=48 km/h	v=96 km/h	v=192 km/h
f_0=225 MHz	10 Hz	20 Hz	40 Hz
f_0=900 MHz	40 Hz	80 Hz	160 Hz
f_0=1500 MHz	67 Hz	133 Hz	267 Hz

The actual Doppler shift of a wave with angle α relative to the vector of the speed of the vehicle is given by

$$f_D = f_{D\,max}\,\cos(\alpha).\tag{2.2}$$

Typically, the received signal is a superposition of many scattered and reflected signals from different directions so that we may speak not of a Doppler shift but of a Doppler spectrum. Figure 2.1 shows an example of a received VHF signal level for a fast-moving car (v = 192 km/h) as a function of the time for a carrier wave of fixed frequency f_0 = 225 MHz.

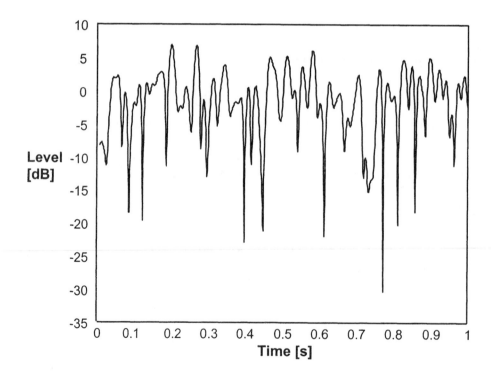

Figure 2.1   Time variance of multipath fading

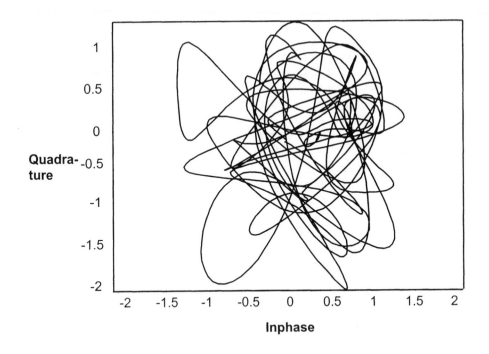

Figure 2.2 Time variance as a curve in the complex plane

The superposition of Doppler-shifted carrier waves leads to a fluctuation of the carrier amplitude *and* the phase. This means the received signal has been amplitude and phase modulated *by the channel*. Figure 2.2 shows the trace of the phasor in the complex plane. For digital phase modulation, these rapid phase fluctuations cause severe problems if the carrier phase changes too much during the time T_S that is needed to transmit one digitally modulated symbol. Amplitude and phase fluctuate randomly. The typical frequency of the variation is of the order of $f_{D\max}$. Consequently, digital transmission with symbol time T_S is only possible if

$$f_{D\max} T_S \ll 1. \tag{2.3}$$

The frequency selectivity of the channel is determined by the different travel times of the signals. They can be calculated as the ratio between the travelling distances and the velocity of light. Table 2.2 shows some typical figures.

Table 2.2 Examples for travel times of the signal

Distance	300 m	3 km	30 km
Time	1 µs	10 µs	100 µs

Travel time differences of some microseconds are typical for cellular mobile radio. For a broadcasting system for a large area, echoes up to 100 µs are possible in a hilly or mountainous region. In so-called single frequency networks (see chapter 7) the system must cope with even longer echoes. Longer echoes correspond to more fades inside the transmission bandwidth. Figure 2.3 shows an example of a received signal level as a function of the frequency at a fixed location where the travel time differences of the signals correspond to several kilometres. In the time domain, intersymbol interference disturbs the transmission if the travel time differences are not much smaller than the symbol duration T_s. A data rate of 200 kbit/s, for example, leads to $T_S = 10$ µs for the QPSK (Quaternary Phase-Shift Keying) modulation. This is of the same order as the echoes. This means that digital transmission of that data rate is not possible without using more sophisticated methods. Known techniques are equalisers, spread spectrum and multicarrier modulation. Equalisers are used in the GSM standard. The data rate for DAB is much higher than for GSM and the echoes in a broadcasting scenario are much longer than in a cellular network. This would lead to a higher complexity for the equaliser. Spread spectrum is spectrally efficient only for cellular networks where it is used as multiple access (CDMA), as in the UMTS standard. For DAB it was therefore decided to use multicarrier modulation, because it is able to cope with very long echoes and it is easy to implement.

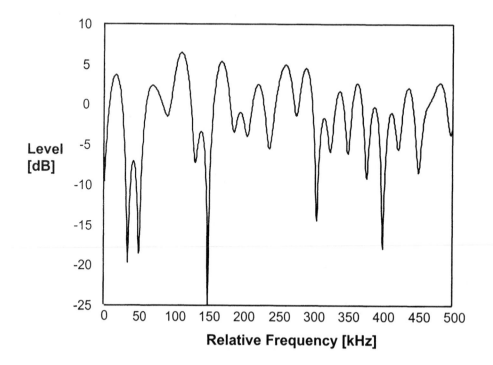

Figure 2.3   Frequency selectivity of multipath fading

For a more detailed treatment of the mobile radio channel and transmissions techniques, we refer to textbooks like (Proakis, 1995), (Kammeyer, 1996) and (David, 1996).

2.2   The DAB Transmission System

2.2.1   Multicarrier Modulation

To cope with the problem of intersymbol interference caused by long echoes, DAB uses a special type of multicarrier modulation: OFDM (Orthogonal Frequency Division Multiplex). The simple idea behind multicarrier modulation is to split up the high-rate data stream into K parallel data streams of low data rate and to modulate each of them separately on its own (sub-)carrier. This leads to an increase of the symbol duration T_S by a factor of K. For sufficiently high K, it is possible to keep T_S significantly longer than the echo duration and to make the system less sensitive to intersymbol interference.

OFDM is a spectrally very efficient kind of multicarrier modulation, because it minimises the frequency separation between the individual carriers by allowing some controlled spectral overlap between the carriers, without causing adjacent channel interference (ACI). This goes back to the mathematical property of orthogonality that gave the name to OFDM.

It is easy to understand an OFDM signal $s(t)$ as a kind of signal synthesis by a finite Fourier series defined by

$$s(t) = \sum_{k=-K/2}^{K/2} z_k \cdot e^{j2\pi kt/T}. \tag{2.4}$$

It is defined on an interval (Fourier period) of length T. The complex Fourier coefficients z_k carry the digitally coded information. For each time interval of length T, another set of $K+1$ information-carrying coefficients can be transmitted. In practice, the DC coefficient for $k = 0$ will not be used (i.e. is set to zero) for reasons of hardware implementation. The Fourier synthesis can be interpreted as a modulation of each complex modulation symbol z_k on a complex carrier wave $\exp(j2\pi kt/T)$ with frequency k/T ($k = \pm1, \pm2,..., \pm K/2$). The signal $s(t)$ is the complex baseband signal and has to be converted to an RF signal by means of a quadrature modulator. At the receiver side, Fourier analysis of the downconverted complex baseband signal will produce the complex symbols using the well-known formula

$$z_k = \frac{1}{T} \int_0^T e^{-j2\pi kt/T} s(t)dt, \tag{2.5}$$

which results from the orthogonality of the carrier waves. Both Fourier analysis and synthesis will be implemented digitally by the FFT (Fast Fourier Transform) and IFFT (Inverse FFT) algorithms. The transmission chain is shown in Figure 2.4.

The part of the OFDM signal that transmits the K complex coefficients z_k is called the OFDM symbol.

To make the transmission more robust against long echoes, the OFDM symbol period T_S will be made longer than the Fourier period T by a so-called cyclic prefix or guard interval of length Δ simply by cyclic continuation of the signal. A synchronisation error smaller than Δ will then only lead to a frequency-dependent but constant phase shift. Echoes are

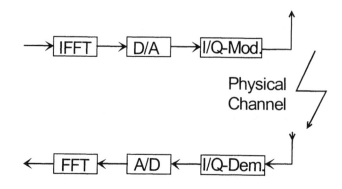

Figure 2.4 FFT implementation of OFDM

superpositions of ill-synchronised signals and will cause no intersymbol interference, but a constant phasor, as long as the delays are smaller than Δ. For DAB, differential quatrature phase shift keying (DQPSK) is used so that this constant phase cancels out at the demodulator.

The length of T_S is limited by the requirement that the phase fluctuations must be small, that is

$$f_{Dmax}T_S \ll 1. \tag{2.6}$$

On the other hand, long echoes require a long guard interval and a long T_S. To keep the system flexible for different physical situations, four Transmission Modes (TMs) with different parameter sets have been defined, see Table 2.3.

Table 2.3 The OFDM parameters for the four DAB transmission modes

Mode	K	$1/T$	T_S	Δ	Max. frequency
TM I	1536	1 kHz	≈1246 μs	≈246 μs	≈375 MHz
TM II	384	4 kHz	≈312 μs	≈62 μs	≈1.5 GHz
TM III	192	8 kHz	≈156 μs	≈31 μs	≈3 GHz
TM IV	768	2 kHz	≈623 μs	≈123 μs	≈750 MHz

The product of the number of sub-carriers K and the spacing $1/T$ between them is the same for all transmission modes and determines the total signal bandwidth of approx. 1.5 MHz. The parameters of all transmission modes can be easily scaled into each other. The ratio Δ/T is always the same. The last column in the table gives a rule of thumb for the maximum transmission frequency due to the phase fluctuation caused by the Doppler effect. A car speed of 120 km/h and a physical channel with no line of sight (the so-called isotropic Rayleigh channel, see (David, 1996)) has been assumed.

Transmission mode I with the very long guard interval of nearly 250 μs has been designed for large-area coverage, where long echoes are possible. It is suited for single frequency networks with long artificial echoes; 200 μs correspond to a distance of 60 km, which is a typical distance between transmitters. If all transmitters of the same coverage area are exactly synchronised and send exactly the same OFDM signal, no signal of relevant level and delay longer than the guard interval will be received. Since the OFDM symbol length T_S is very long, transmission mode I is sensitive against rapid phase fluctuations and should only be used in the VHF region.

Transmission mode II can cope with echoes that are typical of most topographical situations. However, in mountainous regions, problems may occur. This mode is suited for the transmission in the L-band at 1.5 GHz.

Transmission mode III has been designed for satellite transmission. It may be suited also for terrestrial coverage, if no long echoes are expected.

The parameters of TM IV lie just between mode I and II. It was included later in the specification to take into account the special conditions of the broadcasting situation in Canada. It will be used there even at 1.5 GHz. This is possible because of limited speed and direct line of sight.

2.2.2   The Frame Structure of DAB

For each transmission mode, a transmission frame is defined on the physical signal level as a periodically repeating structure of OFDM symbols that fulfil certain tasks for the data stream. It is an important feature of the DAB system (and in contrast to the DVB system) that the time periods on the physical level and on the logical (data) level are matched. The period T_F of the transmission frame is either the same as the audio frame length of 24 ms or an integer multiple of it. As a consequence, the audio data stream does not need its own synchronisation. This ensures a better synchronisation stability especially for mobile reception.

The structure for TM II is the simplest and will thus be described first. The frame length is 24 ms. The first two OFDM symbols of the transmission frame build up the Synchronisation Channel (SC). The next three OFDM symbols carry the data of the Fast Information Channel (FIC) that contains information about the multiplex structure and transmitted programmes. The next 72 OFDM symbols carry the data of the Main Service Channel (MSC). The MSC carries useful information, such as audio data or other services. Figure 2.5 shows the transmission frame structure. It is also valid for TMs I and IV.

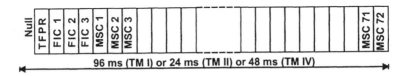

Figure 2.5 Transmission frame structure

All these OFDM symbols in a transmission frame of TM II have the same duration $T_S \approx$ 312 µs, except for the first one. This so-called null symbol of length $T_{Null} \approx 324$ µs is to be used for rough time synchronisation. The signal is set to zero (or nearly to zero) during this time to indicate on the physical level the beginning of a frame. The second OFDM symbol of the SC is called the TFPR (Time–Frequency–Phase Reference) symbol. The complex Fourier coefficients z_k have been chosen in a sophisticated way so that this symbol serves as a frequency reference as well as for channel estimation for the fine tuning of the time synchronisation. Furthermore, it is the start phase for the differential phase modulation. Each of the following OFDM symbols carries 384 DQPSK symbols corresponding to 768 bits (including redundancy for error protection, see below). The three OFDM symbols of the FIC carry 2304 bits. Because they are highly protected with a rate 1/3 code, only 768 data bits remain. The FIC data of each transmission frame can be decoded immediately without reference to the data of other transmission frames, because this most important information must not be delayed. The 72 OFDM symbols of the MSC carry 55296 bits, including error protection. This corresponds to a (gross) data rate of 2.304 Mbit/s. The data capacity of 55296 bits in each 24 ms time period is organised in so-called Capacity Units (CUs) of 64 bits. In the MSC many audio programmes and other useful data services are multiplexed together. Since each of them has its own error protection, it is not possible to define a fixed net data rate of the DAB system.

The transmission frames of TMs I and IV have exactly the same structure. Since the OFDM symbols are longer by a factor of 2 or 4, respectively, the transmission frame length is 48 ms or 96 ms. The number of bits in the FIC and MSC increases by the same factor, but the data rate is always the same.

For TM III, the frame duration is T_F = 24 ms. Eight OFDM symbols carry the FIC, and 144 OFDM symbols carry the MSC. The data rate of the FIC is higher by a factor of 4/3 compared to the other modes. The MSC always has same data rate.

For all four transmission modes, the MSC transports 864 CUs in 24 ms. There is a data frame of 864 CUs = 55296 bits common for all transmission modes that is called the Common Interleaved Frame (CIF). For TMs II and III, there is exactly one CIF inside the transmission frame. For TM I, there are four CIFs inside one transmission frame of 96 ms. Each of them occupies 18 subsequent OFDM symbols of the MSC. The first is located in the first 18 symbols, and so on. For TM IV, there are two CIFs inside one transmission frame of 48 ms. Each of them occupies 36 subsequent OFDM symbols of the MSC.

2.2.3   Channel Coding

The DAB system allows great flexibility in the choice of the proper error protection for different applications and for different physical transmission channels. Using rate compatible punctured convolutional (RCPC) codes introduced by (Hagenauer, 1988), it is possible to use codes of different redundancy without the necessity for different decoders. One has a family of RCPC codes originated by a convolutional code of low rate that is called the mother code. The daughter codes will be generated by omitting specific redundancy bits. This procedure is called puncturing. The receiver must know which bits have been punctured. Only one Viterbi decoder for the mother code is necessary. The mother code used in the DAB system is defined by the generators (133,171,145,133) in octal notation. The encoder is shown as a shift-register diagram in Figure 2.6.

The mother code has the code rate R_c = 1/4, that is for each data bit a_i the encoder produces four coded bits $x_{0,i}$, $x_{1,i}$, $x_{2,i}$ and $x_{3,i}$. As an example, the encoder output corresponding to the first eight data bits may be given by four parallel bit streams written in the following matrix (first bit on the left hand side):

$$
\begin{array}{cccccccc}
1 & 0 & 1 & 1 & 0 & 1 & 1 & 0 \\
1 & 1 & 1 & 1 & 0 & 0 & 1 & 0 \\
1 & 1 & 0 & 0 & 1 & 0 & 1 & 0 \\
1 & 0 & 1 & 1 & 0 & 1 & 1 & 0
\end{array}
$$

A code of rate 1/3 or 1/2, respectively, can be obtained by omitting the last one or two rows of the matrix A code of rate 2/3 (= 8/16) can be obtained by omitting the last two columns and every second bit in the second column If we shade every omitted (punctured) bit, we get the matrix

$$
\begin{array}{cccccccc}
1 & 0 & 1 & 1 & 0 & 1 & 1 & 0 \\
1 & 1 & 1 & 1 & 0 & 0 & 1 & 0 \\
1 & 1 & 0 & 0 & 1 & 0 & 1 & 0 \\
1 & 0 & 1 & 1 & 0 & 1 & 1 & 0
\end{array}
$$

For 8 data bits now only 12 encoded bits will be transmitted· the code has rate 8/12. Using this method, one can generate code rates 8/9, 8/10, 8/11, . 8/31, 8/32 The puncturing pattern can even be changed during the data stream, if the condition of rate compatibility is taken into account (Hagenauer, 1988)

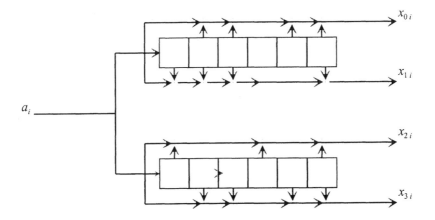

Figure 2.6 Encoder for the DAB mother code

RCPC codes offer the possibility of Unequal Error Protection (UEP) of a data stream some bits in the data stream may require a very low bit error rate (BER), others may be less sensitive against errors Using RCPC codes, it is possible to save capacity and add just as much redundancy as necessary

UEP is especially useful for audio data They are organised in frames of 24 ms The first bits are the header, the bit allocation (BAL) table, and the scale factor select information (SCFSI) An error in this group would make the whole frame useless Thus it is necessary to use a strong (low-rate) code here The next group consists (mainly) of scale factors Errors will cause annoying sounds ("birdies"), but these can be concealed up to a certain amount on the audio level The third group is the least sensitive one It consists of sub-band samples. A last group consists of Programme-associated Data (PAD) and the cyclic redundancy check (CRC) for error detection in the scale factor (of the following frame). This group requires approximately the same protection as the second one. The distribution of the redundancy over the audio frame defines a protection profile. An example is shown in Figure 2.7.

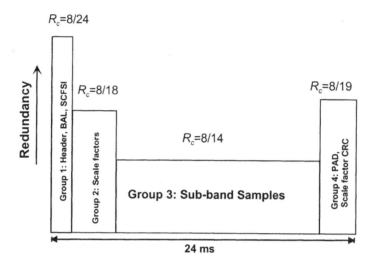

Figure 2.7 Example of an audio UEP profile

The PAD may be extended to the so-called X-PAD. In this case, the PAD group size increases and the audio sub-band sample group decreases. It is important to note that the error protection does not take this into account. The X-PAD is thus worse protected (see section 2.3.3.2).

For audio data with a sampling frequency of 48 kHz, the DAB system allows 14 different data rates between 32 and 384 kbit/s. The protection profiles for all these date rates are grouped into five Protection Levels PL1 to PL5. Inside each protection level different data rates are possible, but the robustness against errors is the same. This means, for example, that if a broadcaster switches between 192 and 256 kbit/s, the audio quality will change, but not the coverage area. PL1 is the most robust protection level, PL5 the least robust one. All protection levels except PL5 are designed for mobile reception; 14 data rates and five protection levels lead to 70 possible combinations. For 64 of them, a protection profile is defined. Table 2.4 shows the possible combinations and the required number of capacity units.

The DAB system allows eight protection levels for Equal Error Protection (EEP). They are intended for data transmission. For the co-called A-profiles 1-A, 2-A, 3-A, 4-A, all data rates are possible that are integer multiples of 8 kbit/s. For the B-profiles the data rate must be a multiple of 32 kbit/s. Table 2.6 shows the eight protection levels and their code rates. The third column shows the number of CUs required for a 64 kbit/s data stream. The fourth column shows the required SNR to reach a BER of $2 \cdot 10^{-4}$ for TM II in a Rayleigh fading channel with $f_{Dmax} = 40$ Hz. The fifth column shows the same for $f_{Dmax} = 125$ Hz.

Table 2.4 Capacity needed for the possible combinations of audio data rates and protection levels

Data Rate	PL1	PL2	PL3	PL3	PL5
32 kbit/s	35 CUs	29 CUs	24 CUs	21 CUs	16 CUs
48 kbit/s	52 CUs	42 CUs	35 CUs	29 CUs	24 CUs
56 kbit/s	X	52 CUs	42 CUs	35 CUs	29 CUs
64 kbit/s	70 CUs	58 CUs	48 CUs	42 CUs	32 CUs
80 kbit/s	84 CUs	70 CUs	58 CUs	52 CUs	40 CUs
96 kbit/s	104 CUs	84 CUs	70 CUs	58 CUs	48 CUs
112 kbit/s	X	104 CUs	84 CUs	70 CUs	58 CUs
128 kbit/s	140 CUs	116 CUs	96 CUs	84 CUs	64 CUs
160 kbit/s	168 CUs	140 CUs	116 CUs	104 CUs	80 CUs
192 kbit/s	208 CUs	168 CUs	140 CUs	116 CUs	96 CUs
224 kbit/s	232 CUs	208 CUs	168 CUs	140 CUs	116 CUs
256 kbit/s	280 CUs	232 Cus	192 CUs	168 CUs	128 CUs
320 kbit/s	X	280 CUs	X	208 CUs	160 CUs
384 kbit/s	416 Cus	X	280 CUs	X	192 CUs

Note It can be seen from the figures in Table 2 4 that the coding strategy supports many possible changes of configuration For example, if a 256 kbit/s audio channel is split up into two 128 kbit/s channels at the same protection level, they will require the same capacity Furthermore, in most cases one can increase the protection to the next better level and lower the audio data rate by one step without changing the required capacity Such a diagonal of constant capacity of 140 CUs has been marked by shading in Table 2 4 It is possible to multiplex several audio channels of different size together, as long as their total size does not exceed 864 CUs Table 2 5 shows as an example the number of 192 kbit/s audio programmes that can be transmitted for the different protection levels and the signal-to-noise ratio (SNR) that is needed at the receiver in a typical (not fast) fading channel (Schulze, 1995) A small capacity for data services is always left

Table 2.5 Number of 192 kbit/s audio programmes and required SNR

Protection Level	Number of Programmes	SNR
PL1	4	7 4 dB
PL2	5	9 0 dB
PL3	6	11 0 dB
PL4	7	12 7 dB
PL5	8	16 5 dB

The protection levels 4-A and 4-B are very sensitive to fast fading They should not be used for mobile applications

All the channel coding is based on a frame structure of 24 ms These frames are called logical frames They are synchronised with the transmission frames, and (for audio) with the audio frames At the beginning of one logical frame the coding starts with the shift registers in the all-zero state At the end, the shift register will be forced back to the all-zero state by appending 6 additional bits (so-called tail bits) to the useful data to help the Viterbi decoder. After encoding such a 24 ms logical frame builds up a punctured codeword. It always contains an integer multiple of 64 bits, that is an integer number of CUs. Whenever necessary, some additional puncturing is done to achieve this. A data stream of subsequent logical frames that is coded independently of other data streams is called a sub-channel. For example, an audio data stream of 192 kbit/s is such a possible sub-channel. A PAD data stream is always only a part of a sub-channel. After the channel encoder, each sub-channel will be time interleaved independently as described in the next subsection. After time interleaving, all sub-channels are multiplexed together into the common interleaved frame (CIF).

Table 2.6   EEP levels: code rate, 64 kbit/s channel size, and required SNR

Protection Level	$R_c=$	Size of 64 kbit/s	SNR (40 Hz)	SNR (125 Hz)
1-A	1/4	96 CUs	5.0 dB	5.4 dB
2-A	3/8	64 CUs	7.1 dB	7.6 dB
1-B	4/9	54 CUs	8.4 dB	8.8 dB
3-A	1/2	48 CUs	9.3 dB	10.0 dB
2-B	4/7	42 CUs	10.6 dB	11.5 dB
3-B	4/6	36 CUs	12.3 dB	13.9 dB
4-A	3/4	32 CUs	15.6 dB	19.0 dB
4-B	4/5	30 CUs	16.2 dB	21.5 dB

Convolutional codes and their decoding by the Viterbi algorithm are treated in textbooks about coding, see for example (Clark, 1988), (Bossert, 1998), (Proakis, 1995). The paper of (Hoeher, 1991b) gives some insight how the channel coding for DAB audio has been developed. It reflects the state of the research work on this topic a few months before the parameters were fixed. Bit error curves of the final DAB coding scheme in a mobile fading channel and a discussion of the limits of the system can be found in the paper of (Schulze, 1995).

2.2.4   Interleaving and PSK Mapping

For an efficient error correction with a convolutional code, a uniform distribution of channel bit errors (before the decoder) is necessary. A mobile radio channel produces burst errors, since many adjacent bits will be disturbed by one deep fade. For OFDM, this holds in time and in the frequency direction. To reach a more uniform distribution of badly received bits in the data stream before the decoder, the encoded bits will be spread over a larger time–frequency area before being passed to the physical channel. This procedure is called (time and frequency) interleaving. At the receiver, this spreading has to be inverted by the deinterleaver to restore the proper order of the bit stream before the decoder.

2.2.5   Time interleaving

To spread the coded bits over a wider time span, a time interleaving will be applied for each sub-channel. It is based on a so-called convolutional interleaver. First, the codeword (i.e. the bits of one logical frame) will be split up into small groups of 16 bits. The bits with number 0 to 15 of each group will be permuted according to the bit reverse law (i.e. 0→0, 1→8, 2→4, 3→12,..., 14→7, 15→15). Then, in each group, bit no. 0 will be transmitted without delay, bit no. 1 will be transmitted with a delay of 24 ms, bit no. 2 will be transmitted with a delay of 2×24 ms, and so on, until bit no. 15 will be transmitted with a delay of 15×24 ms. At the receiver side, the deinterleaver works as follows. In each group bit no. 0 will be delayed by 15×24 ms, bit no. 1 will be belayed by 14×24 ms, and so on, bit no. 14 will be delayed by 24 ms, and bit number 15 will not be delayed. Afterwards, the bit reverse permutation will be inverted. Obviously, the deinterleaver restores the bit stream in the proper order, but the whole interleaving and deinterleaving procedure results in an overall decoding delay of 15×24 ms = 360 ms. This is a price that has to be paid for a better distribution of errors. A burst error on the physical channel will be broken up by the deinterleaver, because a long burst of adjacent (unreliable) bits before the deinterleaver will be broken up so that 2 bits of a burst have a distance of at least 16 after the deinterleaver and before the decoder.

The time interleaving will only be applied to the data of the MSC. The FIC has to be decoded without delay and will therefore only be frequency interleaved.

2.2.6   DQPSK Modulation and Frequency Interleaving

Because the fading amplitudes of adjacent OFDM sub-carriers are highly correlated, the modulated complex symbols will be interleaved. This will be done with the QPSK symbols before interleaving. We explain it by an example for TM II: a block of 768 encoded bits have to be mapped onto the 384 complex coefficients for one OFDM symbol of duration T_S. The first 384 bits will be mapped to the real parts of the 384 QPSK symbols, the last 384 bits will be mapped to the imaginary parts. To write it down formally, the bits of the lth block $p_{i,l}$ ($i = 0, 1,..., 2K-1$) will be mapped to the QPSK symbols $q_{i,l}$ ($i = 0,1,..., K-1$) according to the rule

$$q_{i,l} = \frac{1}{\sqrt{2}}\left[(1 - 2p_{i,l}) + j(1 - 2p_{i+K,l})\right], \quad i = 0,1,...K-1. \tag{2.7}$$

The frequency interleaver is simply a renumbering of the QPSK symbols according to a fixed pseudo-random permutation $F(i)$, as shown in Table 2.7. The QPSK symbols after renumbering are denoted by $y_{k,l}$ ($k = \pm1, \pm2, \pm3,..., \pm K/2$).

Table 2.7   Permutation for frequency interleaving (TM II)

i	0	1	2	3	4	5	...	380	381	382	383
$k=F(i)$	−129	−14	−55	−76	163	141	...	−116	155	94	−187

The frequency interleaved QPSK symbols will be differentially modulated according to the law

$$z_{k,l} = z_{k,l-1} \cdot y_{k,l}. \tag{2.8}$$

The complex numbers $z_{k,l}$ are the Fourier coefficients of the OFDM symbol no. l in the frame.

2.2.7   Performance Considerations

Sufficient interleaving is indispensable for a coded system in a mobile radio channel. Error bursts during deep fades will cause the Viterbi decoder to fail. OFDM is very well suited for coded transmission over fading channels because it allows time and frequency interleaving. Both interleaving mechanisms work together. An efficient interleaving requires some incoherency of the channel to get uncorrelated or weakly correlated errors at the input of the Viterbi decoder. This is in contrast to the requirement of the demodulation. A fast channel makes the time interleaving more efficient, but causes degradations due to fast phase fluctuations. The benefit of time interleaving is very small for $f_{Dmax} < 40$ Hz. On the other hand, this is already the upper limit for the DQPSK demodulation for TM I. For even lower Doppler frequencies corresponding to moderate or low car speeds and VHF transmission, the time interleaving does not help very much. In this case, the performance can be saved by an efficient frequency interleaving. Long echoes ensure efficient frequency interleaving. As a consequence, SFNs support the frequency interleaving mechanism. If, on the other hand, the channel is slowly *and* frequency flat fading, severe degradations may occur even for a seemingly sufficient reception power level. A more detailed discussion of these items can be found in the paper of (Schulze, 1995).

2.3   The DAB Multiplex

2.3.1   Mode-Independent Description of the Multiplex

The DAB system is designed for broadcasting to mobiles in the frequency range from 30 MHz to 3 GHz. This cannot be achieved by a single OFDM parameter set, so four different transmission modes are defined (see section 2.2.1). The DAB multiplex, however, can be described independently of the transmission mode. To achieve this, containers of information are defined which are used to transmit the data of applications (audio and data services, service information, etc.) to the receivers. Figure 2.8 shows the generation of the DAB multiplex.

The data of audio components and other applications are carried in what is called the Main Service Channel (MSC). Every 24 ms the data of all applications are gathered in sequences, called Common Interleaved Frames (CIFs). Multiplex and service-related information is mainly carried in the Fast Information Channel (FIC). Similar to the MSC, FIC data are combined into Fast Information Blocks (FIBs).

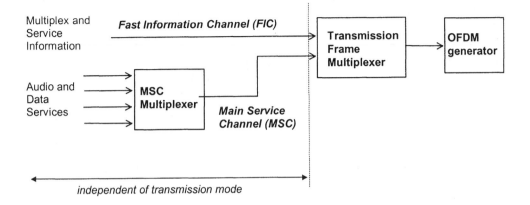

Figure 2.8 Generation of the DAB multiplex

Depending on the transmission mode, a number of CIFs and FIBs are grouped together into one transmission frame which is the mapped to a number of OFDM symbols (see section 2.2.2).

2.3.2   The Main Service Channel

The MSC of the DAB system has a gross capacity of 2.304 Mbit/s. Depending on the convolutional code rate, the net bit rate ranges from approximately 0.6 to 1.8 Mbit/s. Single applications do not normally consume this overall capacity. The MSC is therefore divided into sub-channels. Data carried in a sub-channel are convolutionally encoded and time interleaved. Figure 2.9 shows the conceptual multiplexing scheme of the DAB system.

The code rate can differ from one application to another. The data rates available for individual sub-channels are given by integer multiples of 8 kbit/s (of 32 kbit/s for some protection schemes). Figure 2.10 shows an example for a multiplex configuration. Each sub-channel can be organised in stream mode or packet mode.

The division of the MSC into sub-channels and their individual coding profiles are referred to as the Multiplex Configuration. The configuration is not fixed but may be different for different DAB transmissions or may vary from time to time for the same transmission. Therefore the multiplex configuration must be signalled to the receivers. This is done through the FIC.

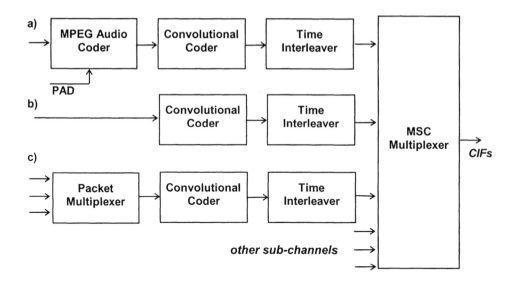

Figure 2.9 Generation of the common interleaved frames (CIFs) from a) an audio sub-channel,
b) a general data stream mode sub-channel, c) a packet mode sub-channel.

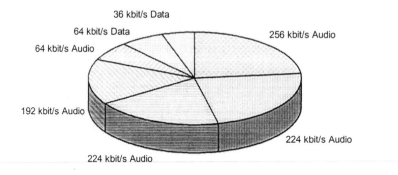

Figure 2.10 An example of how the overall capacity is assigned to individual applications (capacity
pie).

2.3.3   Transport Mechanisms

The DAB system provides several mechanisms to transport data to receivers which are each tailored to specific purposes.

The Stream Mode is designed for continuous and synchronous data streams such as coded audio. In this case every 24 ms the same number of bits is transmitted transparently, that is, there is no mechanism inside the DAB system to provide synchronisation or addressing except by the frame structure. All the necessary signalling has to be provided by the application itself. Stream mode sub-channels may contain audio encoded according to ISO 11172-3 Layer II or general data at fixed bit rate.

For asynchronous data there is Packet Mode, which provides a protocol for conveying single data groups through a packetised channel. The packet protocol allows repetition of data to be handled and the creation of a multiplex of several parallel applications, to which the capacity can flexibly assigned.

A special way of transport is provided for Programme-associated Data (PAD) which are inserted into the MPEG Layer II audio data stream by defining a structure of the Auxiliary Data field of the MPEG audio frame specific to DAB. It provides a number of functions related to the contents of the audio programme and can be inserted at the place where the audio is produced. Therefore PAD is considered to be a part of the audio and not really a separate transport mechanism for data.

2.3.3.1   Stream Mode

Stream mode is used for applications which can provide a constant data rate of a multiple of 8 kbit/s (32 kbit/s for the B coding profiles, see section 2.2.3). For example, at a sampling rate of 48 kbit/s, the MPEG Layer II audio encoder generates a data frame every 24 ms which exactly meets this requirement. When transmitting general data, the data stream can be divided into "logical frames" containing the data corresponding to a time interval of 24 ms. These logical frames can be transmitted one after the other in the same manner as MPEG audio frames.

When stream mode is used there are two options for error protection. Unequal error protection (UEP) is used with MPEG audio and provides error correction capabilities which are tailored to the sensitivity of the audio frame to bit errors (see section 2.2.3). For general data, equal error protection (EEP) is used, where all bits are protected in the same way.

2.3.3.2   Programme-associated Data

Although DAB provides mechanisms to carry general data in packet mode, there is a need for an additional method for transmitting data which is closely linked to an audio service. This is referred to as PAD.

At the end of the MPEG audio frame there is an auxiliary data field which is not further specified in MPEG. For the DAB audio frame, however, this field is used to carry the CRC for the scale factors (see Chapter 3), and two PAD fields, see Figure 2.11. The DAB system is transparent for the PAD. This means that the PAD can be produced and inserted at the time when the audio signal is coded, normally at the studio (see Chapters 5 and 6) and will be retrieved only when decoding the audio signal in the receiver.

The fixed PAD (F-PAD) field enjoys the same protection level as the SCF-CRC field and hence is well protected. Therefore, it can be used to signal Dynamic Range Control (DRC, see section 5.4) and other control information to the receiver.

End of DAB audio frame

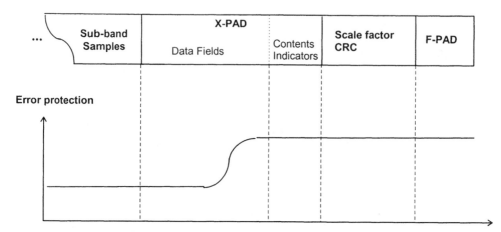

Figure 2.11 The PAD fields at the end of the DAB audio frame

The extended PAD (X-PAD) field can be used to send larger amounts of data (up to 64 kbit/s), for example text messages or multimedia objects (see Chapter 4). To create a flexible multiplex structure within the PAD, a special packet structure was developed. The data are arranged in X-PAD-Data Groups which each consist of a data field and a Contents Indicator which signals the kind of data carried in the corresponding data field and in some cases the number of bytes of the data field. Data fields of several applications can be carried in parallel in the PAD section of one audio frame.

The extended PAD (X-PAD) field partly enjoys the protection level of the SCF-CRC; the larger part, however, only enjoys the protection level of the sub-band samples. Therefore the contents indicators are separated from the data fields and collectively sent in the better protected part to ensure that the important information on the contents and amount of data can be received with high probability.

2.3.3.3   Packet Mode

The most general DAB transport mechanism is the packet mode structure.

For packet mode transmission, the data is organised in data groups which consist of a header, a data field of up to 8191 bytes and optionally a cyclic redundancy check (CRC) for error detection. The data group header allows identification of different data group types such that, for instance, scrambled data and the parameters to access them can be carried in the same packet stream. There are also counters which signal if and how often the data group will be repeated. An extension field offers the possibility to address end user terminals or user groups. Data groups are transmitted in one or several packets.

The packets themselves constitute the logical frames in packet mode similar to the audio frames in stream mode. To fit into the DAB structure requiring a multiple of 8 kbit/s, that is a multiple of 24 bytes per 24 ms, packet lengths of 24, 48, 72 and 96 bytes are defined in the DAB standard. The packets consist of a 5 byte header, a data field, padding if necessary, and a CRC for error detection. The packets may be arranged in any order in the transmission. The header of each packet signals its length so that the decoder can find the next packet. An important feature of packet mode is that padding packets can be inserted if no useful data are available to fill the data capacity. Packet mode is therefore suitable for carrying asynchronous data. Up to 1023 applications may be multiplexed in a packet mode transmission.

Figure 2.12 shows an example of a packet mode transmission. Two applications are carried in parallel. From each application, a suitable amount of data, for example one file, is selected to form a data group. The corresponding header is added to the data and the CRC is calculated. Each data group is mapped on to a number of packets, each of which may have a different length. The first, intermediate and last packets carrying data of one data group are marked accordingly in their headers. The packets of different applications use different packet addresses. Therefore, the sequence of the packets belonging to the first application may be interrupted by those of the second application. However, within each sequence of packets, the transmission order has to be maintained to allow the decoder to correctly reassemble the data group.

For more details of the data group and packet headers the reader is referred to (EN 300401) and (TR 101496).

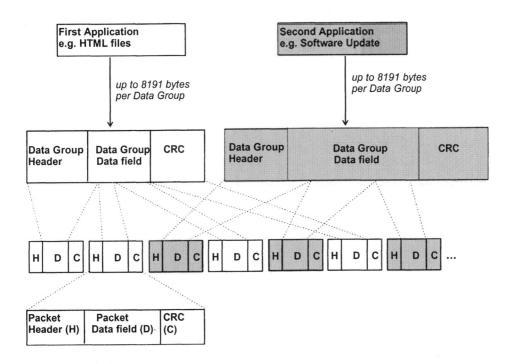

Figure 2.12 Example of a packet mode transmission multiplexing two applications

2.3.4   Fast Information Channel

The FIC is used to signal the multiplex configuration of the DAB transmission. Therefore, it uses fixed symbols in the transmission frame which are known to the receivers. The receiver needs this information to be able to decode any of the sub-channels. Therefore, for instant data acquisition, the FIC data are not time interleaved (hence the name "fast"). Instead, a high protection (code rate 1/3) and frequent repetition of the data is used to guarantee high availability.

The FIC consists of a number of Fast Information Blocks (FIBs) which carry the information (see Figure 2.13). Each FIB is made up from 32 bytes: 2 CRC bytes are used for error detection, the remaining 30 bytes are filled with Fast Information Groups (FIGs) in which the Multiplex Configuration Information (MCI) and other information are coded. The FIGs are distinguished by a type field. FIGs of type 0 are used for the MCI and Service Information (SI), FIGs of type 1 are used to send text labels, FIGs of type 5 are used to send general data in the FIC (Fast Information Data Channel, FIDC), FIGs of type 6 are used with access control systems and FIGs of type 7 are used for in-house data transmission by the broadcaster. Some FIG types use extensions to indicate different meanings of the data fields, e.g. FIG type 0 extension 1 is used to signal the sub-channel organisation while FIG type 0 extension 10 is used to send the date and time.

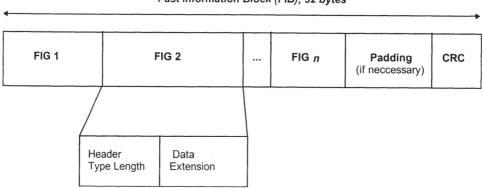

Figure 2.13   Structure of a Fast Information Block (FIB) and a Fast Information Group (FIG)

In special cases the data rate provided by the FIB symbols in the transmission frame (32 kbit/s) may not be sufficient to carry all of the FIC information with the desired repetition rate. In this case, part of this information can be sent in parallel in packet mode sub-channel #63 with packet address 1023. This is referred to as the Auxiliary Information Channel (AIC). The corresponding packet mode data group contains the FIGs as defined for the FIC. Data concerning other DAB transmissions may be sent only in the AIC. For more details see section 2.5.2.4.

2.3.5   Transmission Frames

From FIBs and CIFs transmission frames are formed which are then mapped to OFDM symbols. Table 2.8 lists the duration of the frames and the number of FIBs and CIFs used in each transmission mode.

Table 2.8   The frames of the four DAB transmission modes

Mode	Duration of Transmission Frame, ms	Number of CIFs per Transmission Frame	Number of FIBs per Transmission Frame
TM I	96	4	12
TM II	24	1	4
TM III	24	1	3
TM IV	48	2	6

2.3.6   The logical structure of the DAB Multiplex

The different data streams carried in the DAB multiplex can be grouped together to form what is called a Service. The services can be labelled (e.g. "Bayern 3") and it is the services from which the listener makes his or her choice. All services taken together are referred to as an Ensemble.

The different data streams (e.g. audio, data, etc.) which belong to one service are called its Service Components. Different services may share components, and the logical structure of services, service components and the position in the CIF where the data of the components are actually carried is signalled as part of the MCI in the FIC. Figure 2.14 shows an example of this structure. The ensemble labelled DAB 1 contains a number of services, among them Radio 11, Radio 12, Data 29 and Radio 3. Radio 3 is a service which consists of an audio component only, that is a "normal" radio programme. Radio 11 and Radio 12 each consist of an audio component and share Radio 1 data as a common data component which carries information relevant to both programmes or programme-independent data such as traffic messages. At times, say during a news programme on the hour, Radio 11 and Radio 12 broadcast the same audio signal. Instead of transmitting the corresponding bits twice, it is possible to signal this at the level of service components. The capacity in the CIF usually used for the audio component of Radio 12 (i.e. Sub-ch. 7) can be reused during this time for another service or component, for example slow motion video. Data 29 is a data service without an audio component but with two separate data components. Figure 2.14 also shows how the different components are transported in the MSC. It can be seen that different components (Radio 1 data and Data 29 data 2) may share a packet mode sub-channel while stream mode components each require an individual sub-channel.

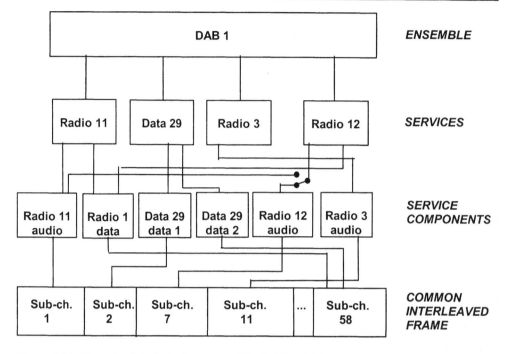

Figure 2.14 Example of the logical structure of the DAB multiplex

It is also apparent in the figure which information has to be sent in the FIC: the sizes and position of the sub-channels in the CIF and their respective code rates are signalled as "Sub-channel organisation" in FIG type 0 extension 1, the services and their components are described by "Basic service and service component definition" in FIG type 0 extension 2 and 3 (used for packet mode).

Of course there is a need to provide further information about each service, such as service label, programme type, programme number for recorder control, indication of announcements (e.g. for automatically switching to a service broadcasting traffic information), alternative frequencies and frequencies of other DAB transmissions. This kind of information may be displayed in order to assist the listener in operating the receiver or to enable the receiver to optimise the reception. The DAB system provides a number of sophisticated mechanisms for this purpose, far beyond what has been realised by the Radio Data System (RDS) in FM broadcasting (see section 2.5).

2.3.7   Multiplex Reconfiguration

To achieve a high flexibility, the multiplex may be reconfigured from time to time during transmission. When the multiplex configuration is about to change, the new information, together with the timing of the change, is transported via the MCI and indicates in advance what changes of the multiplex are going to take place. Because of the time interleaving, special precaution has to be taken when the size of a sub-channel changes. For details the reader is referred to Annex D of (EN 300401).

2.4   Conditional Access

Access control may be an important feature especially for data applications on DAB. But also scrambled audio services could be required in some broadcasting environment. To serve these needs, an access control mechanism is included in the DAB specifications.

The access control system consists of three major parts:
- scrambling / descrambling,
- entitlement checking,
- entitlement management.

By using these means, a number of features can be realised, such as subscription per theme, level or class, pay per programme, per service or time. The only part of the system which is specified in the DAB standard (EN 300401) is the scrambling and descrambling process. For entitlement checking and management different systems, including proprietary ones, may be used.

2.4.1   Scrambling and Descrambling the Bits

In the scrambling system which is used by DAB, the individual bits are scrambled by adding (using the logical AND function, or adding "modulo 2") a pseudo-random binary sequence (PRBS) to the data. The PRBS is derived from an Initialisation Word (IW) fed into a PRBS generator, which consists of a number of coupled shift-register stages and is often implemented as an integrated circuit. The PRBS generator used in DAB is the same as the one used for scrambling TV signals and is specified in (EN 300174).

A receiver with the PRBS generator implemented can therefore descramble the received signal and hence access the service if it can generate the correct IWs at all times. The IWs change very often to restart the PRBS generator. This avoids long delays when the service is accessed and ensures that only short PRBSs are generated from which the IW cannot be derived. This is important for broadcast applications because the PRBS generator is publicly known and hence a pirate could acquire an entitlement and calculate the PRBS by comparing the scrambled and unscrambled signals.

To allow for frequently changing IWs they are generated from two sources: the Control Word (CW), which is provided by the entitlement checking entity (see below), and the Initialisation Modifier, which is either generated locally from the DAB frame count or transmitted along with the data in plain (i.e. unscrambled).

In stream mode (audio and data), the complete stream is scrambled, and the PRBS generator is initialised at the beginning of each logical frame. In packet mode and in the FIDC only the data fields of the packets or FIGs are scrambled, not the headers and CRCs. In packet mode it is also possible to scramble an entire data group before distributing the data to different packets.

2.4.2   Checking Entitlements

To be able to check if a receiver is entitled to unscramble a service, the service provider sends an Entitlement Checking Message (ECM) to the access control systems in the receivers. This message contains the CW scrambled by a second security system, which is kept secret and in which the different access control systems differ, and the conditions which the user has to fulfil to be able to unscramble the CW. In the access control system in the receiver these parameters are stored, for example on a smart card which is inserted in the receiver and which the user has purchased and paid for. The ECM is descrambled in the access control system and the parameters sent by the service provider are compared with those stored locally. If they match, the CW and hence the IW can be calculated.

Depending on the transport mechanisms, several ways of transmitting the ECMs are possible.

In stream mode (audio or data) the ECMs have to be sent in a separate channel because there is no means of multiplexing the scrambled data and the ECMs inside the sub-channel. The ECMs are therefore transported in the FIC (FIG type 6) or in sub-channel #63 using packet mode.

In packet mode, it is possible to send both the scrambled data and the ECMs in the same sub-channel, because they can be distinguished by their data group headers. This makes it possible to store a service component sent in packet mode (e.g. on a computer disk) and descramble off-line using software only instead of a PRBS circuit.

2.4.3   Entitlement Management

The entitlement management function makes it possible to update the access parameters stored in the receivers. This may be used to extend a subscription over the air without the need to send a new smart card to the subscriber.

For this purpose the service provider sends Entitlement Management Messages (EMMs) to individual decoders, which therefore need to have a unique address. For certain purposes it is also necessary to address groups of decoders or all decoders. The EMMs are also carried in packet mode data groups or in the FIC.

2.5   Service Information

2.5.1   Introduction

Service Information (SI) provides additional information about the services carried in an ensemble and is intended to simplify service access and to provide attractive receiver features and functionality. Similar information carried in other ensembles and/or in AM/FM services may also be signalled. Unlike the MCI (Multiplex Configuration Information), SI features are optional, that is broadcasters and receiver manufacturers may choose which to implement and when.

2.5.2   Basic Information

2.5.2.1   **Ensemble Label**

The listener basically finds a programme by selecting a service in an ensemble.

The ensemble label coded in FIG type 1 Extension 0 contains the Ensemble Identifier (EId) and a character field formed by a string of 16 characters which provide the listener with a clear textual description for the ensemble. A character flag field contains a flag for each character in the ensemble label. When a bit is set to "1", the corresponding character of the ensemble label is included in an abbreviated label for display on receivers having a display of less than 16 characters.

Example:

Ensemble label 16 characters :	`__BBC_National__`
Character flag field :	`0011111110000100`
Ensemble label 8 characters :	`BBC_Natl`

2.5.2.2   **Date and Time**

The date and time information in DAB is encoded in MJD (Modified Julian Date) and UTC (Co-ordinated Universal Time). The MJD is a continuous count of the number of days elapsed since 17 November 1858. The UTC is a 24 hour atomic time system that forms the basis of most radio time systems.

The data and time feature is encoded in FIG type 0 extension 10 which contains besides the *MJD* and the *UTC* also a leap second indicator (LSI) and a confidence indicator. The ensemble provider should set the leap second indicator to "1" throughout a UTC day that contains a leap second. The confidence indicator is set to "1" when the time information is within an agreed tolerance. The MJD is coded as a 17-bit binary number that increments daily at 0000 (UTC) and extends over the range 0-99999. The following formula allows the MJD to be calculated from the year (Y), month (M) and date (D):

MJD = 14956 + D + int [(Y-L) · 365.25 + int [M + 1 + L·12)·30.6001]. For January and February (i.e. for M = 1 or M = 2) variable L = 1, otherwise L = 0.

Example : 10 October 2004 -> MJD = 14956 + 10 + int [(104-0)· 365.25 + int [10 + 1 + 0·12)·30.6001] = 69715 = 00100010001000001010011 = 0x111053.

FIG 0/10 allows the time information to be provided at two different resolution levels. The standard resolution allows the time to be expressed in 1 minute intervals (short form of FIG 0/10) whilst the high-resolution level allows timing to be expressed in millisecond intervals (long form of FIG 0/10). Five bits are used to represent the hour (in the range 0 – 23) and 6 bits are used to represent the minutes. The long-form version contains additionally a 6-bit "seconds" and a 10-bit "milliseconds" sub-field.

Example: 10 October 2004 at 14h 45min 25sec 50 ms:

MJD	Hours	Minutes	Seconds	Milliseconds
00100010001000001010011	01110	101101	0110001	0000110010
(69715= 10 Oct 2004)	(14)	(45)	(25)	(50)

2.5.2.3   Country, Local Time Offset, International Table

A number of basic, mostly static, ensemble parameters are encoded in FIG type 0 extension 9 (FIG 0/9) :

- The Ensemble ECC (Extended Country Code) specifies together with the country code of the EId a world-wide unique ensemble identifier. In an international ensemble it is possible to assign different ECCs to services. This is done in the extended field. The ECC is crucial in case of PTy preview or PTy downloading (see section 2.5.4.2) and also for TMC (see Chapter 4).

- The Ensemble LTO defines the local time offset of the ensemble from the UTC transmitted in FIG type 0 extension 10 (see section 2.5.2.2). It is expressed in multiples of half an hour (in the range −12 hours to +12 hours). In an international ensemble it is possible that some services use a different LTO. This can be signalled in the extended field.

- The Inter. Table Id indicates which international table is applicable in the ensemble. Two international tables are defined:
  - Table 00000001 comprises the European table of RDS PTY codes (except for codes 30 and 31) as specified in CENELEC EN 50067 and the table of announcement types specified in ETSI EN 300401.
  - Table 00000010 comprises the RBDS list of PTY codes as specified in the RBDS standard and the table of announcement types specified in ETSI EN 300401.

The coding allows the definition of future table identifiers, for example for use in Asian or African countries where there might be a need for another list of PTys and/or another list of announcement types. The use of different international tables in one ensemble is not allowed. In Europe only Table 00000001 is applicable.

2.5.2.4   FIC Redirection

When there is a risk for overload in the FIC some FIGs may be transferred either as a whole or in part to the Auxiliary Information Channel (AIC). The AIC uses packet mode transport. FIC redirection is encoded in FIG type 0 extension 31. The FIG type 0 flag field and the FIG type 1 flag field indicate which FIGs of type 0 resp. type 1 are carried in the AIC. All MCI FIGs should remain in the FIC. FIGs applying to other ensembles may be entirely transferred to the AIC. Also FIG type 1 extension 2 (PTy downloading) and FIG type 0 extension 30 (satellite database) may be entirely transferred to the AIC. Other FIGs may be present partly in the FIC and partly in the AIC. For some of them the information is repeated more frequently in the FIC than the AIC or vice-versa. This ensures that those receivers not equipped with a packet mode decoder will be able to access the basic data features though their performance may be inferior to other receivers.

2.5.3   *Service-related Information*

2.5.3.1   Programme Service Label and Data Service Label

The service label feature comprises a character field forming a string of up to 16 characters and provides the listener with a textual description of the service name. The characters are coded from byte 15 to byte 0. The first character starts at byte 15. The character field is linked to a machine-readable Service Identifier (SId). In the case of a programme service label the SId comprises 16 bits and is similar in format to the RDS Programme Identification (PI) code: the 4 MSBs form the Country Id which distinguishes between fifteen countries while the remaining 12 bits are the service reference. In the case of a data service label the SId comprises 32 bits: the 8 MSBs represent the service ECC, the next 4 bits the Country Id and the remaining 20 bits the service reference. The character flag field allows identification of up to eight characters from the service label to suit receivers with a short display. If the *character field* contains less than 16 characters, the unused bits in the character flag field (having no corresponding character) are set to zero as shown in Table 2.9.

Table 2.9   Example of service labels

	Programme Service Label	Data Service Label
FIG	FIG 1/1	FIG 1/5
SId	16 bits	32 bits
	4 bits Country Id	8 bits service ECC
	12 bits service reference	4 bits Country Id
		20 bits service Reference
Example of service label: 16 characters	_ _BBC_R1_Digital_	_ _RADIO_1_TMC_ _ _
character flag field	01111111100000000	011 0 0 000111 1 1 100
Service label 8 characters	_BBC_R1_	_ R1_TMC_

Whereas the ensemble label (see section 2.5.2.1) is expected to change very rarely, if ever, the name of the service will probably change only when the programme is continued by a different service provider, which may happen a few times per day, if ever.

2.5.3.2   Service Component Label

Service component labels are useful for the user to distinguish between service components of the same type. All service components are accessed at the service level, which means that service components can only be requested after their parent service has been selected. The receiver is required to support an additional layer of service access: firstly to offer the services and secondly to offer the components of a chosen service. The service component label is encoded in FIG type 1 extension 4 in terms of the SCIds, which identifies the service component within the service.

Service component labels may change frequently. It is recommended that the receiver follows any changes to these labels to avoid obsolete information remaining in the display.

Example: Service "RADIO 1" with SId = F123 (hex) and service label "RADIO 1" has in addition to the primary service component two secondary (audio) service components labelled as "RADIO 1 EXTRA" and "RADIO 1 PLUS":

Service	Service Label	SCIdS	Service Component Label
F123	RADIO 1	0001	RADIO_1_ EXTRA
		0010	RADIO_1_ PLUS

2.5.3.3   Service Component Trigger

The beginning or the end of the transmission of a service component can be signalled by means of the Service Component Trigger (SCT) feature. It is primarily used with data services or multimedia services. In the case of receivers processing the FIC or the AIC in a continuous way, the SCT may be transmitted in the FIC or in the AIC at any time in order to signal the beginning or end of the transmission of a service component. The same SCT may be transmitted several times without any constraint on the transmission time. When a receiver is a member of a service user group and it satisfies the access conditions, it shall look for the service carried in the MSC in stream or packet mode or in the FIDC, at the time indicated by the SCT. Since there may be a delay between the real time of the reception of the service (component) and the time indicated in the SCT the SCT should be transmitted by enough time in advance. For the start of the service, the actual beginning of the reception of the service (component) shall be later than the time indicated in the SCT. For the end of the service (component), when signalled, the end time indicated shall be later than the actual end time.

The SCT feature is encoded in FIG type 0 extension 20 (FIG 0/20) which comprises the service component description in terms of the service component type (ASCTy or DSCTy). This FIG also signals where the component is located: in the FIDC or in the MSC in stream mode (short form of FIG 0/20) or in the MSC in packet mode (long form of FIG 0/20). FIG 0/20 further contains a number of flags. One of the flags signals whether the Logical Frame Number (LFN) applies to the beginning or end of service transmission. Another flag is used to signal whether the optional service user group information (defining specific groups of receivers) is present or not. Additional information may be provided to signal the time (UTC in 1 minute resolution) and conditional access parameters (CAId, SCCA).

If the time value is set to "0", or is absent, and if LFN = "1FFF", the transmission of the indicated service (component) is considered to have just started or ended.

2.5.3.4   Programme Number

The Programme Number (PNum) is a code that enables receivers and recorders designed to make use of this feature to respond to the particular programme item(s) that the user has pre-selected. In a receiver PNum may be implemented in a similar way as PTy in WATCH mode (see also section 2.5.4.2). However, whereas PTy WATCH is used to switch on or record a particular genre of programme, PNum allows a specific programme to be switched on or recorded.

PNum uses the scheduled programme time of the programme, to which the day of the month is added in order to avoid ambiguity. PNum together with the SId of the service that has scheduled the programme item form a unique identifier for this programme item. PNum is similar to the 16-bit PIN (Programme Item Number) in RDS. The first 5 MSBs (b15 to b11) represent the day of the month, in the range 00–31, the next 5 bits (b10 to b6) the hour, in the range 00–23, and the final 6 bits (b5 to b0) the minute, in the range 00–59. When the date part (b15 to b11) of PNum = 00000 special codes are allowed for the hours and minutes to signal a status code, a blank code or an interrupt code as shown in Table 2.10.

PNum is encoded in FIG type 0 extension 16 where it is associated with the SId to provide a unique reference for each programme. The Continuation flag indicates whether or not there will be a planned interruption while the Update flag indicates a re-direction to a different service (New SId) and time (New PNum). New SId thus indicates the SId of the target service of the redirection while New PNum specifies the new emission time or the time at which the programme will be resumed. It is coded in the same way as PNum.

Table 2.10   Special PNum codes

Date	Hours	Minutes	Description	Meaning
b15–b11	b10–b6	b5–b0		
00000	00000	000000	Status code	no meaningful PNum is provided
00000	00000	000001	Blank code	not worth recording (e.g. a non-scheduled programme during an intermission)
00000	00000	000010	Interrupt code	unplanned interruption of a programme, which will continue after a short interval (e.g. during a traffic announcement)

Example 1: Redirection of a programme item

Suppose Radio 1 (SId = F123) has scheduled on the 18th of the month, at 16.25h, a programme item called "Festival d'Avignon". At 17.00h the news is scheduled so that "Festival d'Avignon" is continued on Radio 3 (SId = F456) where it replaces a scheduled programme item "Today in Paris" which is deleted.

The PNum for "Festival d'Avignon" has D = 18 = 10010; H = 16 = 10000; M = 25 = 011001 so that PNum = 1001 0100 0001 1001 = 0x9419.

At 16.25h the following FIG 0/16 is transmitted:

SId	PNum = 9419			Cont	Upd
	Date	Hour	Min		
F123	18	16	25	0	0

Receivers programmed to start recording "Festival d'Avignon" on Radio 1 can start recording this programme item. These receivers have to be informed that this programme item will continue at 17.00h on Radio 3. Since "Festival d'Avignon" is redirected to service F456 at 16.59h, the parameters for the New PNum are: D = 18 = 10010; H = 16 = 10000; M = 59 = 111011 so that New PNum = 1001 0100 0011 1011 = 0x943B.

At 16.57h the following FIG 0/16 is transmitted during at least the last minute of the programme:

SId	PNum = 9419			Cont	Upd	New SId	New PNum = 943B		
	Date	Hour	Min				Date	Hour	Min
F123	18	16	25	1	1	F456	18	16	59

The receiver is now informed that the programme item with PNum = 9419 (D18-H16-M25), namely "Festival d'Avignon", continues on F456 (Radio 3) under a new PNum = 943B (D18-H16-M25). However, on Radio 3 a programme item was scheduled at 17.00h with PNum = 9440 (D18-H17-M00) called "Today in Paris". This programme will be deleted.

When a programme is deleted, the original SId and PNum codes are transmitted with the Update flag set to "1" and with New PNum = blank code. When a new programme item replaces the original scheduled programme item the PNum code for the new programme item is set to a new unique start time which is equal to the nominal start time of the original programme item minus 1 minute.

Therefore at 17.00h, the following two FIGs 0/16 are transmitted in parallel:

SId	PNum = 943B			Cont	Upd
	Date	Hour	Min		
F456	18	16	59	0	0

SId	PNum = 9440			Cont	Upd	New SId	New PNum = 0001		
	Date	Hour	Min				Date	Hour	Min
F456	18	17	00	0	1	F456	00	00	01

With the first FIG 0/16 receivers are informed that at 17.00h on F456 (Radio 3) a (new) programme item with PNum = 943B (D18-H16-M59), that is "Festival d'Avignon", replaces the original scheduled programme item.

The second FIG 0/16 signals that the originally scheduled programme item with PNum = 9440 (D18-H17-M00), that is "Today in Paris", is deleted. The New PNum in this FIG is a blank code indicating that it is not worth recording this programme item.

This second FIG 0/16 (which is interleaved with the first FIG 0/16) is transmitted until at least 17.01h.

Example 2: Unplanned interruption
During a traffic announcement (i.e. an unplanned interruption) the Update and Continuation flags are set to "0". Suppose Radio 1 (SId = F123) has scheduled on the 18th of the month at 9.15h a programme entitled "The Renaissance" and at 9.30h a programme item entitled "Festival d'Avignon".

At 9.15h a first FIG 0/16 is transmitted (1) with PNum = 924F. Receivers are informed that this is the PNum of "The Renaissance". Receivers programmed to record this programme item can start recording. At 9.18h there is a traffic announcement. A recorder recording "The Renaissance" should be warned to stop the recording temporarily. Therefore a new FIG 0/16 is transmitted (2) containing the interrupt code (PNum = 0002). When the traffic announcement is over, the same FIG 0/16 is transmitted (3) as in case (1) to allow the receiver to resume recording of "The Renaissance". At 9.30h programme item "Festival d'Avignon" starts. A new FIG 0/16 is transmitted (4) to allow receivers programmed to record this item to start recording while receivers programmed to record "The Renaissance" will stop recording.

FIG 0/16	SId	Date	Hour	Min	PNum	Cont	Upd	Remarks
(1)	F123	18	09	15	924F	0	0	"The Renaissance"
(2)	F123	00	00	02	0002	0	0	"Traffic announcement"
(3)	F123	18	09	15	924F	0	0	"The Renaissance"
(4)	F123	18	09	30	925E	0	0	"Festival d'Avignon"

Example 3: Delay of a programme item
If the start of a programme item is delayed due to the extension of a previous programme item a FIG 0/16 with Update flag = "1" and containing a New PNum with the new start time is transmitted. Suppose Radio 1 (SId = F123) has scheduled on 18th May, at 15.05h, a programme item called "Roland Garros" with PNum (D18-H15-M05) = 93C5 and at 15.45h a programme item called "Today in Paris" with PNum = D18-H15-M45 = 93ED. Because of the extension of "Roland Garros" the start of "Today in Paris" is postponed by 10 minutes until 15.55h.

At 15.45h the following FIG 0/16 is transmitted:

SId	PNum = 93C5			Cont	Upd
	Date	Hour	Min		
F123	18	15	05	0	0

This FIG 0/16 informs the receiver that at 15.45h not the expected "Today in Paris" but a programme item with PNum = 93C5, i.e. "Roland Garros", is (still) on air. In other words "Roland Garros" is extended until after 15.45h. Receivers programmed to record "Today in Paris" with PNum = 93ED should be informed that this programme item is postponed until 15.55h. Therefore another FIG 0/16 having the update flag = 1 and containing as New PNum = D18-H15-M55 = 93F5 (indicating the new start time of "Today in Paris") is interleaved with the former FIG 0/16.

SId	PNum = 93ED			Cont	Upd	New SId	New PNum = 93F7		
	Date	Hour	Min				Date	Hour	Min
F123	18	15	45	0	1	F123	18	15	55

Receivers programmed to record "Today in Paris" will wait until 15.55h to start recording.

2.5.4   *Programme-related Features*

2.5.4.1   Service Component Language

The Service Component Language (SC Lang) feature signals the language associated with the content of a service component. It is used for service (component) selection based on language rather than the SId, which is required for conventional service selection.

SC Lang serves two purposes:

1. In the case of audio service components, SC Lang is used to identify the spoken language associated with the audio content of a service component. SC Lang is required whenever a programme service has more than one secondary service component in a different language. If a service has no secondary service components (only a primary service component) then PTy language in FIG 0/17 (see 2.5.4.2) is used to signal the language of the programme. In this case SC Lang provides complementary redundant information that can be used for confirmation purposes.
2. In the case of data service components, SC Lang is used to identify the language associated with the content of the data. It allows the user selection of data service components by dint of language.

SC Lang is encoded in FIG type 0 extension 5 (FIG 0/5) where the language is associated with either the SubchId, the FIDCId or the SCId depending on whether the component is carried in the MSC in stream mode, in the FIC or in the MSC in packet mode. In the first case the short form of FIG 0/5 is used (L/S flag = "0"), in the latter case the long form is used (L/S flag = "1"). The MSC/FIC flag indicates whether, in the short form, the component is carried in the MSC in stream mode or in the FIC.

In the following example the SC Lang for a service component carried in the MSC in stream mode is Swedish in sub-channel "3" and Finnish in sub-channel "4".

L/S	MSC/FIC flag	SubchId	Language
0	0 (MSC)	000011 (3)	00101000 (0x 28) = Swedish
0	0 (MSC)	000100 (4)	00100111 (0x 27) = Finnish

SC language versus PTy language

It is important to notice the subtle difference between SC language and PTy language. SC Lang can be used to signal the language of any service component within an ensemble, being primary or secondary, audio or data. PTy language is intended to be used in combination with the programme types and it signals the language of the primary service component of a service. In addition, one language can be signalled for the secondary service component(s). Whereas SC language cannot be used to signal the language of service components in another ensemble, PTy language can be cross-referenced to another ensemble.

Whenever both SC language and PTy language are implemented their language fields should be consistent, that is, SC language should be equal to PTy language. If all services within an ensemble are in the same language, there is basically no need for signalling SC language. However, its use is strongly recommended because some receivers may have implemented a language filter based on SC language. In the case of static PTy language, the SC language should also be static. When the service provider has a service with programmes targeted to different language audiences then it should signal the language dynamically. If it has no means for dynamic signalling, static signalling is allowed on condition that the PTy language is set equal to "0". SC language in this case is not transmitted.

2.5.4.2   Programme Type

The PTy feature provides another means of service access based on the categorisation of programme content by the service provider and the transmission of codes representing these categories to the receiver. The PTy classification in DAB has been greatly extended beyond the 32 potential categories available with RDS. More than one code can be assigned at a time. This reduces the number of times that code "0" (no PTy) need be used and allows programmes to be more accurately described.

Coarse and Fine PTy codes

There are two levels of PTy categorisation referred to as coarse and fine. Besides the 30 categories which are available from the international table (and which provide full compatibility with RDS in Europe or RBDS in the United States), a further table of 32 additional coarse codes (32–64) is available to extend the categories of the first table.

Examples of possible coarse codes: Opera, Musical, Agriculture, Politics, Tourist interests, Consumer interests, Elderly interests, Women's interests, Disabled interests, Minority/Ethnic, Foreign language, Environmental, Wildlife/Nature, etc.

A further set of up to 256 fine codes is available for finer sub-divisions of some of the categories defined in the combined pair of tables of coarse codes. Fine codes not only provide a more detailed description of the programme but also allow a more efficient WATCH mode, especially in the case of PTy preview. It makes more sense to look for TENNIS than for SPORT or to look for VIOLIN CONCERTO than for CLASSICS. Table 2.11 shows examples of possible fine codes (together with the associated international code or coarse code). Each fine code may be associated with only one course code.

Table 2.11   Examples of possible PTy fine codes

Int. code or Coarse code	Possible PTy Fine Codes
SPORT	Athletics, Winter sports, Water sports, Equestrian sports, Martial arts, Non-instrumental ball games, Instrumental ball games, Football, Tennis, Wimbledon, Cricket, Basketball, Hockey, Formula 1, Baseball, Skiing, etc.
CLASSICS	Baroque music, Chamber music, Contemporary music, Vocal & choral music, Ballet music, Live concert, Recital, Cantata, Symphony, Piano concerto, Mozart, etc.
POP	Reggae, Beat, Techno, House, Disco, Rapp, Funk, The Sixties, etc.
JAZZ	Blues, Ragtime, New Orleans, Dixieland, Spiritual, Swing, etc.
OPERA	Grand Opera, Comic Opera, Belcanto, Oratorio, Operetta, Zarzuela, Singspiel, etc.

It is expected that the several coarse codes will be agreed internationally. Others will be defined only at a national level. Fine codes may be different for each country but some of them may also be agreed internationally.

For coarse codes in the range 32 to 63 and all fine codes, the PTy downloading mechanism (see section 2.5.4.3) provides the means for their "over-air" definition by linking these codes to a displayable label.

Static and Dynamic PTy codes

Service providers may wish to implement the PTy feature in different ways. The first way minimises the expense of the studio infrastructure by keeping the codes relatively static. This approach serves to reflect the general service flavour, that is PTy codes (as well as the language, if signalled) are assigned to a service and remain unchanged for a long period and may be used to SEARCH for specific programme categories. It is obvious that in this case the PTy may not reflect the actual on-air programme but rather the format or genre of the service as a whole. Secondly, the PTy feature may be operated in a dynamic mode. In this case the code changes at programme item junctions for example from NEWS to SPORT or from NEWS in SWEDISH to NEWS in FINNISH. Dynamic PTy allows the receiver to operate in the WATCH mode as well as in SEARCH or SELECTION mode. Theoretically up to eight PTy codes can be allocated to a programme: 1 static Int.code, 1 static coarse code, 2 static fine codes, 1 dynamic Int.code, 1 dynamic coarse code, 2 dynamic fine codes.

The Int.code, Coarse code and Fine code fields

PTy is encoded in FIG type 0 extension 17 (FIG 0/17) where the SId of a service is linked to a language field, an Int.code, one coarse code and up to two fine codes.

Example: Suppose there is an international agreement for the following coarse codes: 32 = OPERA, 33 = AGRICULTURE; and for the following fine codes: 108 = TENNIS and 176 = FLAMENCO MUSIC.

Signalling code "14" (CLASSICS) in the Int.code field of FIG 0/17 and code "32" in the coarse code field can signal an OPERA. The fine code field can contain a nationally agreed fine code, for example "147" = "Salzburger Festspiele".

A programme on AGRICULTURE can be signalled by putting value "02" (current affairs) or "03" (information) or "00" (no programme type) in the Int.code field of FIG 0/17 and value "33" in the coarse code field.

A report from a TENNIS match can be signalled by putting value "04" (SPORT) in the Int.code field and value "108" in the fine code field. In this case the coarse code field is not used but a second (nationally agreed) fine code could be used, for example, 124 = ROLAND GARROS.

Putting code "26" (National music) in the Int.code field and code "176" in the fine code field can signal a programme with FLAMENCO MUSIC. Also in this case the *coarse code* field can but need not be used. It would for example be possible to put code "15" (Other music) in the Int. code field and code 26 (national music) in the coarse code field (or vice versa) as shown in the example below.

SId	S/D	Language	Int.code	Coarse Code	Fine Code
A301	1	00001000 German	01110 (14) Classics	100000 (32) Opera	10010011 (147) Salzburger Festspiele
8217	1	00011101 Dutch	00011 (03) Information	100001 (33) Agriculture	Absent
F238	1	00001111 French	00100 (04) Sport	Absent	01101100 (108) Tennis 01111100 (124) Roland Garros
E265	1	00001010 Spanish	01111 (15) Other music	11010 (26) Nat. music	10110000 (176) Flamenco music

PTy of primary and secondary service components

The PTy categorisation applies to the primary audio service component of a service. Secondary service components, if available, cannot be individually categorised. They should be of the same "type" as the primary component but they may or may not be given in the same language. One language code can be assigned to a service, regardless of how many secondary components there are. The language therefore applies to the primary component, which is signalled by the *P/S flag* in FIG 0/17 being set to "primary". There is one situation where the language of a secondary component can be signalled. This is when the service comprises only two audio components, one primary and one secondary, and these two components have different languages.

Example: Suppose a Swedish service E456 has on-air a news programme in Swedish (as primary SC) and in Finnish (as secondary SC). This can be signalled by means of two FIGs 0/17:

SId	S/D flag	P/S flag	L flag	CC flag	Language	Int.code
E456	1 (dynamic)	0 (primary)	1	0	00101000 Swedish	00001 News
E456	1 (dynamic)	1 (secondary)	1	0	00100111 Finnish	00001 News

It is important to notice that when a service comprises secondary service components a SEARCH or WATCH will basically yield the primary service component.

Example: Service F123 with PTy code = SPORT has a primary service component transmitting a report of a TENNIS match and at the same time a secondary service component transmitting a report of a FOOTBALL match. A SEEK or a WATCH for SPORT will yield the report of the TENNIS match. If a service contains one (and not more than one) secondary service component in a different language a WATCH for the (common) PTy will yield the component in the desired language if the receiver has a language filter based on PTy language. In the example above of the Swedish service E456, a WATCH for NEWS in FINNISH will directly yield the Finnish news on the secondary service component of that service.

2.5.4.3   Programme Type Downloading

The PTy downloading feature allows a receiver to be updated with PTy descriptions that do not belong to the international table of fixed codes or to other codes that have been internationally agreed. PTy downloading applies to coarse and fine codes and is also used to establish a link between a coarse code and a maximum of two fine codes.

PTy downloading is encoded in FIG type 1 extension 2 (FIG 1/2) that relates the (coarse or fine) PTy code to a 16-character label describing the programme type. The language field indicates the language of the PTy label.

The example below illustrates the definition of fine code "57" (Piano music), in German, by means of FIG 1/2. The fine code is linked to coarse code "Classics" and labelled "KLAVIERMUSIK" in the Character field. The language is "German" and the code is valid in Germany (CountryId = "D", ECC = "E0").

Coarse code	Fine code	Language	Character field	Chr-flag field	E C C	Country Id
01110 (14)	00111001 (57)	00001000 (German)	_ _KLAVIERMUSIK_ _	0111111110 000000 KLAVIER	E0	D

The PTy downloading feature can also be used to transmit definitions of the fixed international codes in different languages. Internationally and most nationally defined codes are seldom redefined. However, a number of national codes may be set aside for special events and may occasionally be redefined. Once a downloaded code has expired this code should not be reused for a sufficiently long period, in order not to confuse receivers in which this code may have been pre-set.

2.5.4.4   Programme Type Preview

The PTy preview feature allows ensemble and service providers, which implement dynamic PTy codes, to provide a preview of (coarse or fine) PTy codes for programmes which are planned to be broadcast in the next one or two hours, in the tuned ensemble as well as in another ensemble. The PTy preview feature is attractive for listeners operating the PTy feature in WATCH mode as they can now be assured that the PTy code they have selected will indeed become available.

PTy preview is encoded in FIG type 0 extension 12 (FIG 0/12). A flag field containing coarse codes (in the range 0 to 31 and/or 32 to 63) or fine codes (in the range 0 to 127 or 128 to 255) is linked to the EId (Ensemble Identifier) and the language.

The example below illustrates a FIG 0/12 signalling PTy preview in Ensemble F123 where some services will start soon with programmes having PTy coarse codes NEWS, SPORT, CLASSICS and POP.

EId	L flag	FF	Language	Flag field
F123	1	00 (coarse codes 0 to 31)	00001111 French	00000000000000000100010000010010 (classics, pop, sport, news)

The language field in FIG 0/12 should be consistent with the service component language of FIG 0/5. The flags in the Flag field which signal the PTy coarse or fine codes should be reset to "0" as soon as the programme to which the code is allocated is actually transmitted.

It is important to notice that national PTy codes can have a different meaning in different countries, for example, coarse code "48" could mean AGRICULTURE in Germany while in France coarse code "48" could mean WOMEN's INTEREST. PTy preview applies to an ensemble, not to a service, that is, it is unknown in advance which service will start (soon) with the programme having the previewed code. For an unambiguous interpretation of a national PTy code both the Country Id and the ECC have to be known. An international ensemble may contain one or more services having a Country Id and ECC different from the rest of the services in the ensemble. However, the ECC of another ensemble cannot be signalled from the current ensemble (FIG 0/9 does not feature Other Ensemble functionality). Only by temporarily tuning to the other ensemble will the ECC of that ensemble be known. The consequence is that in some cases a preview of national PTy codes (coarse or fine) can be ambiguous and should therefore not be done.

2.5.4.5   Announcements
The announcement feature is similar to the traffic announcement feature of RDS but on a larger scale, providing more categories such as news flashes or weather bulletins. Sixteen announcement categories can be coded of which eleven are currently defined.

Interrupt mechanism
Just like TA in RDS (see section 5.4) the announcement feature in DAB allows a listener to be directed from the current service/audio source to a programme service, which delivers a short audio message.

It is important to notice the subtle difference between the announcement feature and vectored dynamic PTys. "News announcements" for example are basically short flashes, dynamic and unscheduled for short vectored interrupts. PTy "News" on the other hand denotes a "type of programme" which mostly lasts tens of minutes and which is scheduled in advance according to a programme number (see also section 2.5.3.4).

The DAB announcement interruption mechanism is dependent on the following filter criteria:

- **The type of announcement,** i.e. is the type of announcement supported by the service and has the user selected it?

  For every service which is allowed to be interrupted by announcements, static announcement support information is provided in FIG type 0 extension 18 (FIG 0/18), which comprises 16 announcement support (Asu) flags. These indicate the announcement types (supported within the current ensemble) which can interrupt the service. This information may be used by the listener to select those announcement types for which he or she wants the audio source to be interrupted and to deselect those for which he or she doesn't want an interruption. The announcement switching information in FIG 0/19 comprises an announcement switching (Asw) flags field indicating the type of ongoing announcement; the SubChId to identify the sub-channel that generates the announcement, the cluster Id to identify the cluster to which the announcement is directed and optionally the region for which the announcement is applicable. The Asu flags are independent of the cluster Id. A special implementation of the announcement feature employs a dedicated announcement channel occupying its own sub-channel.

- **The cluster Id,** i.e. does the service belong to the cluster to which the announcement is directed?

  A cluster represents a group of services to which an announcement is directed with the intention of interrupting all the services participating in the cluster. A cluster Id can therefore be considered as "a permission-to-interrupt" indicator. The announcement support information of FIG 0/18 includes a list of cluster Ids, which identify those announcement clusters a service is participating in. If an announcement is sent to a cluster that is not in the list, it will not interrupt the service. Cluster Id 0000 0000 specifies that an ongoing announcement is intended to interrupt the (audio) service components of all services having the signalled sub-channel identifier in their service organisation information. Cluster Id 1111 1111 is used for alarm announcements. The alarm announcement deviates from the other announcement types in that cluster Id 11111111 causes the receiver to interrupt all services in the ensemble.

- **The Region Id** (optional), i.e. does the service belong to the region to which the announcement is targeted and has the user selected this region?

  When an announcement is targeted to a particular geographical region, the Region Id is appended to the announcement switching information of FIG 0/19. When the region Id is absent the announcement is targeted to the whole ensemble service area. In the receiver a region filter is considered to be a valuable tool that can allow listeners to select information about any region in the ensemble service area and not just in the area where the receiver is situated.

- **The New flag,** i.e. is the announcement that is being broadcast a new message or not?

  The New flag, signalled in the announcement switching information, is used by the service provider to distinguish between a message that is being broadcast for the first time ("new") and a message that is repeated ("old"). The New flag is suited for a cyclic announcement channel, that is a sub-channel reserved for repeated emission of announcements and possibly not belonging to a specific service. The detection of the flag by the receiver allows the listener (if desired) to avoid being interrupted by repeated messages announcements.

The cluster concept

Suppose a service provider is operating five services grouped in following three clusters:

Cluster Id 0000 0111 (7)	Cluster Id 0000 1000 (8)	Cluster Id 0000 1001 (9)
F301 (Radio 1)	F301 (Radio 1)	F302 (Radio 2)
F302 (Radio 2)	F303 (Radio 3)	F304 (Radio 4)
F303 (Radio 3)		

Table 2.12 shows an example of possible announcement support information for these services by means of several FIGs 0/18.

Table 2.12   Example of announcement support information

SId	Asu Flags	No. of Clusters	Cluster Id 1	Cluster Id 2
F301	0000010000110010	00010	00000111 (7)	00001000 (8)
Radio 1	Traffic, News, Weather, Finance	(2)		
F302	0000000001000000	00010	00000111 (7)	00001001 (9)
Radio 2	Event	(2)		
F303	000001000001000	00010	00000111 (7)	00001000 (8)
Radio 3	Finance	(2)		
F304	0000001000000010	00001	00001001 (9)	
Radio 4	Sport, Traffic	(1)		

Assume that at a certain moment Radio 1 generates an area weather flash. Radio 1 is participating in two clusters: 7 and 8. If the service provider does not want to disturb the listeners of Radio 2 then it will direct, in the announcement switching information of FIG 0/19, the weather flash to cluster "8" so that only Radio 3 listeners may be interrupted.

Cluster Id	Asw flags	New flag	SubchId	RegionId (lower part)
00001000 (8)	0000000000100000 Weather flash	1	011011 (27)	100010 (34) Bretagne

This FIG 0/19 signals that the ongoing announcement is a new "weather" announcement, generated in sub-channel "27", directed to cluster "8" and targeted to region "34" (Bretagne). This weather flash will interrupt service F303 if the user has not deselected it and if either the receiver is located in Bretagne or the user has selected "Bretagne" from the region menu.

It is clear that the cluster concept is a powerful tool for the service provider to control the interruption of services by the various announcements generated by the services in the multiplex. In multilingual ensembles for example the cluster concept can be used to prevent listeners from being interrupted by announcements in a language they don't understand.

Other ensemble announcement feature

Other ensemble (OE) announcement support information is encoded in FIG type 0 extension 25 (FIG 0/25) and switching information in FIG type 0 extension 26 (FIG 0/26). The latter is used to signal the identity of the ensemble providing the announcement, together with the Cluster Id Current Ensemble and the Cluster Id Other Ensemble.

In the same manner as for the basic announcement feature, the receiver can determine whether an interruption is permitted. The frequency information (see section 2.5.5.1) together with the EId Other Ensemble allows the receiver to determine the appropriate centre frequency of the other ensemble. The receiver can then retune to that frequency and decode the basic announcement switching information (FIG 0/19) in the other ensemble.

Consider several other ensembles with overlapping service areas and covering an area within the service area of the ensemble to which the receiver is tuned. When an announcement is ready to be broadcast in one of the other ensembles, the announcement switching information (FIG 0/19) for that ensemble is relayed to the provider of the tuned ensemble so that the appropriate OE service announcement switching information (FIG 0/26) can be generated.

The region Id specified within the switching information allows the ensemble provider to derive and supply one or more appropriate region identifiers from knowledge of the service area of the OE service. The region Ids must be chosen from those defined for use within the tuned ensemble. When region Ids are not allocated for the current ensemble, Region Id Current Ensemble = 000000 must be used. When region Ids are not allocated for the other ensemble the region flag must be set to "0" or Region Id Other ensemble "000000" must be used.

Example of OE announcement support (FIG 0/25):

SId	Asu flags	No. of EIds	EId 1	EId2
F123	0000000000010000 Area weather flash	0010 (2)	F444	F555

This FIG 0/25 signals that in two other ensembles (F444 and F555) "area weather flash" announcements are supported which can possibly interrupt service F123. It is important to notice that only one announcement support flag can be set at a time. For each other supported announcement type a separate FIG 0/25 must be transmitted.

Example of OE announcement switching (FIG 0/26):

Cluster Id CE	Asw flags	Region Id CE	EId OE	Cluster Id OE	Region Id OE
00000010 (2)	0000000001000000 Sport report	100010 (34) Bretagne	F555	00000111 (7)	001110(14) Rennes

This FIG 0/26 signals that a "sport report" announcement is now ongoing in the other ensemble F555 where it is directed to cluster "7". This cluster corresponds to cluster "2" in the current ensemble, so that the announcement can possibly interrupt service F123. In ensemble F555 the event announcement is targeted to region "14" (Rennes) which maps to region "34" (Bretagne) in the current ensemble.

FM announcements feature

FM announcement support is encoded in FIG type 0 extension 27 (FIG 0/27). FM announcement switching is encoded in FIG type 0 extension 28 (FIG 0/28).

Consider an RDS service covering an area within the DAB ensemble coverage area. When a traffic announcement (TA) is ready to be broadcast on the RDS service, the "TA" flag should be relayed to the DAB ensemble provider so that the appropriate FM service announcement switching information (FIG 0/28) can be generated. Great care has to be taken to synchronise the information that is transmitted at the start of the message.

Therefore a raised TA flag must be present in the RDS service as long as the announcement is going on. The ensemble provider can supply one or more appropriate region Ids from knowledge of the service area of the RDS service. The region Ids must be chosen from those defined for use within the tuned ensemble. The selected region allows the receiver to filter out only those RDS traffic announcements relevant to that region. However, RDS services may not be available over the whole DAB ensemble coverage area. The receiver must therefore check as a background task, using the frequency information feature (see section 2.5.5.1), which RDS services are currently available at the receiver's location. Instead of the listener choosing a particular region(s), the relevant "local" region may be established automatically by using an accurate receiver location-finding mechanism based on the TII feature (see sections 2.5.5.3 and 7.3.7) and regional identification features (see section 2.5.5.4).

Example of FM announcement support (FIG 0/27):

SId	Nr of PI codes	PI1	PI2	PI3	PI4
F123	0100 (4)	FA11	FB22	FC33	FD44

This FIG 0/27 signals that four RDS stations, FA11, FB22, FC33 and FD44, are TP stations on which traffic announcements can be generated which can possibly interrupt DAB service F123 in the current ensemble.

Example of FM announcement switching (FIG 0/28):

ClusterId CE	RegionId CE	PI
00000001 (1)	001110(14) Rennes	FA33

This FIG 0/28 signals that an ongoing traffic announcement on RDS station FA33 (Radio Rennes) is relayed by the DAB ensemble provider to the current ensemble and directed to cluster "1" so that it can possibly interrupt DAB service F123. As the traffic announcement from Radio Rennes is rather of local importance, the DAB ensemble provider has decided to target it to region "Rennes" rather than region "Bretagne".

2.5.4.6   Other Ensemble Services

The Other Ensemble Services (OE services) feature indicates which other ensembles are carrying a particular service. The OE services feature can be used for both programme services and data services. The feature is encoded in FIG type 0 extension 24 (FIG 0/24) where a link between the SId of a service and EIds of other ensembles carrying that service is provided.

When the OE flag in the header of FIG 0/24 header is set to "0", FIG 0/24 provides a link between a service carried in the current ensemble and other ensembles carrying that service. When the OE flag is set to "1", FIG 0/24 provides a link between a service carried in another ensemble and other ensembles carrying that service.

The example below illustrates the case where a programme service F123 (16 bits) is an unscrambled service which is carried in three other ensembles F444, F555, F666.

SId	Rfa	CAId	Number of EIds	EId1	EId2	EId3
F123	0	000	0011	F444	F555	F666

Although the OE services feature is optional it becomes essential if the frequency information associated with a particular service needs to be signalled. However, a receiver, using the OE information alone in order to present the listener with a list of possible services from which to make a selection would produce a long list for the whole country, the majority of which may not be available to the listener. Therefore the Transmitter Identification Information (TII) feature (see section 2.5.5.3) is important because receivers using TII codes in order to identify the transmitter(s) they are tuned to can estimate their own location. OE information should be kept up to date as far as possible. After a multiplex reconfiguration in another ensemble there may be some obsolete information sent for some time. This period should be minimised.

2.5.5 Tuning Aids

2.5.5.1 Frequency Information

The Frequency Information (FI) feature is used to signal the radio frequencies for other DAB ensembles and other broadcast services, such as FM (with and without RDS) and AM.

When used in combination with the region definition feature (see section 2.5.5.4) a geographical area filter may be provided to allow the receiver to determine which of the (many) frequencies listed are worth checking in the area where it is situated.

The FI feature allows mobile receivers leaving the ensemble coverage area to retune to an alternative frequency (service following). The alternative frequency may apply to an identical DAB ensemble (same EId), another ensemble carrying the equivalent primary service component or an RDS or AM service. The services can be identical (SId = PI or dummy code) or hard-linked (see section 2.5.5.2). If there is no alternative source for the same component the FI feature can help receivers to find a primary service component, which is soft-linked to the currently selected one.

The FI feature also helps receivers to get faster access to other DAB ensembles that are available in certain regions. An ensemble scan may be speeded up if the receiver knows the centre frequencies of all other DAB ensembles.

In combination with the OE/FM announcement feature (see section 2.5.4.5) the FI feature can establish a link between services and frequencies needed for announcement switching. In conjunction with other ensembles or the FM PTy WATCH function FI allows fast access to services starting to broadcast a programme item of the desired type.

FI Lists per Region Id

The FI feature is encoded in FIG type 0 extension 21 (FIG 0/21) where, per region Id, one or more FI lists are given. An FI list can contain up to two DAB ensemble frequencies and up to eight RDS frequencies so that a maximum of four DAB ensemble frequencies or 17 RDS frequencies can be signalled per region in one FIG 0/21.

Thanks to the RegionId the receiver can check only a restricted and yet directly relevant list of alternative frequencies but it requires the receiver to know where it is located. The

RegionId is the full 11-bit version so that the region can but need not be a labelled region that is known to the user.

RegionId = "0" means that no area is specified. In this case the FI is valid for the whole coverage area of the signalled ensemble. The R&M field allows differentiating between DAB ensembles and FM services (with or without RDS) and AM services (MW in steps of 9 kHz or 5 kHz).

Continuity Flag

The continuity flag is an indication whether "continuous output" can be expected or not, for example, when switching to another DAB ensemble or when switching to equivalent and hard-linked services. The continuity flag may be set to "1" if the conditions listed in Table 2.13 are fulfilled.

The frequency of an ensemble is denoted as geographically adjacent when this ensemble is receivable somewhere in the region specified in the region Id field. This is important in the case of a multiple frequency network (MFN) comprising several SFNs at different frequencies. Not-geographically-adjacent-area means that it is not guaranteed that the ensemble for which the centre frequency is signalled is receivable somewhere in the region specified in the region Id field.

Table 2.13   Conditions for putting continuity flag = "1"

In case of frequencies of DAB ensembles (covering adjacent areas)	In the control field the transmission mode must be signalled.
	When the frequency of an ensemble is signalled, this ensemble must be synchronised in time and frequency with the current (tuned) ensemble. This means that the null symbols of corresponding transmission frames must coincide (tolerance = guard interval duration). The centre frequency of the signalled ensemble must not deviate by more than 10 Hz from the nominal CEPT frequency.
	Services present in the tuned and referenced ensemble should have identical CU start address, sub-channel size and error protection profile for their primary service component. For the primary service component the CA conditions need to be unique and the CIF counter values must be equal.
	These continuity requirements apply to all frequencies in the "Freq. list" field.
In case of frequencies of FM /AM services	The time delay between the audio signal of the DAB service and the analogue service must not exceed 100 ms.

Importance of the OE flag

Particular attention should be given to the OE flag in the sub-header of FIG 0/21.

* The OE flag is set to "0" if the FI applies to the whole tuned ensemble or to an FM or AM service carrying a primary service component from the tuned ensemble (PI or dummy code = SId).
* The OE flag is set to "1" if the FI applies to frequencies of ensembles other than the tuned ensemble or to FM/AM services which are not identical to a primary service component from the tuned ensemble.

For the latter two cases the continuity flag is set to "0".

Table 2.14 below shows the settings of the OE flag and Cont. flag in relation to the Id field.

Table 2.14   FIG 0/21: OE and continuity flag settings in relation to the Id field

Id Field	R&M	OE Flag	Cont. Flag	Control Field
EId of current ensemble	0000	0	0 or 1	used
EId of another ensemble	0000	1	0 or 1	used
RDS station with PI = SId	1000	0	0 or 1	absent
RDS station with PI ≠ SId	1000	1	0	absent
FM station with dummy Id = SId	1001	0	0 or 1	absent
FM station with dummy Id ≠ SId	1001	1	0	absent
AM station with dummy Id = SId	1010	0	0 or 1	absent
AM station with dummy Id ≠ SId	1010	1	0	absent

Examples of FIG 0/21

Example 1: The following FIG 0/21 illustrates the signalling of the centre frequency of a DAB ensemble F888 in the region of Marseille. The OE flag is set to "1". The ensemble operates in transmission mode II and is receivable somewhere in the Marseille region so that the control field is coded as 00100, meaning "geographically adjacent, transmission mode II".

OE flag	RegionId	Id field	R&M	Cont. flag	Length of freq. list	Control field	Freq1
1	0x028 Marseille	0xF888 (EId)	0000 (DAB)	0	011 (3 bytes)	00100 geog. adj. mode II	0x1C53C (1463.232 MHz)

Example 2: The following FIG 0/21 illustrates the signalling of three RDS frequencies of France Inter for the region of Marseille.

RegionId	Id field	R & M	Cont. flag	Length of freq. list	Freq1	Freq2	Freq3
0x028 Marseille	0xF201 (PI) Fr. Inter	1000 (RDS)	0	011 (3 bytes)	0x26 91.3 MHz	0x2A 91.7 MHz	0x4F 95.4 MHz

2.5.5.2   Service Linking

When different services carry the same audio programme, service linking information is used to signal when the services may be treated as identical; in this case the services are said to be "hard-linked" together. The second use allows services to be generically related.

This means that they do not necessarily carry the same audio but the kind of programmes supported is similar in some way, for example different local services from the same service provider. In this case the services are said to be "soft-linked" together.

When the listener is moving outside the reception area of an ensemble, the service linking feature helps the mobile receiver to find alternatives if the originally tuned service is no longer available. The listener may in this way "follow" and retain the same programme. In the case of dual DAB/RDS receivers the feature allows service following between the same service on DAB and on RDS. In the case of simulcast programmes in different ensembles or simulcasting on DAB and RDS common identifiers may be used instead of service linking because in this case frequency information provides all the information needed for service following.

Short and Long form of FIG 0/6
Service linking is encoded in FIG type 0 extension 6 (FIG 0/6) which has two versions: a short form and a long form.

The short form is used for rapid delinking or to signal a Change Event (CEI) and contains the Linkage Set Number (LSN) together with a number of flags such as the LA flag (Linkage Actuator), the S/H flag (Soft/Hard) and the ILS (International Linkage Set Indicator).

The long form has an Id list of service identifiers appended (PI codes or dummy codes referring to the linked services).

The LSN is a 12-bit code, unique to each set of services that are actually or potentially carrying a common programme.

The LA flag indicates if a service is actually (LA = 1) or potentially (LA = 0) a member of the set of programme services described by the LSN.

The ILS flag indicates that the link applies to only one country (national) or several countries (international).

The combination of the LSN, ILS and S/H parameters identifies a particular set of services and constitutes the "key" for the service linking feature. This key must be unique for the area where it can be received and also for any adjacent area. There is no relation between sets with a common LSN, but with different settings of the ILS. When a link ceases the LSN may be reused again for another set of services.

The IdLQ (Identifier List Qualifier) in the Id list usage field is used to indicate whether the Id in the Id list is a DAB SId or an RDS PI code or a dummy code.

The Shd flag is a shorthand indicator used to indicate that the identifiers in the Id list, having the second nibble in the range "4" to "F", each represents a list of up to 12 services each sharing the same country Id and the same 8 LSBs of the service reference.

Reference Identifier
If one of the services of a linkage set belongs to the current ensemble the OE flag in the FIG 0/6 header is set to "0". The identifier of this service is a "reference identifier" and must be included at the beginning of the Id list regardless of the IdLQ.

Example: Assume a set of 16 DAB services with SId = F211, F212,..., F227 and a set of 16 RDS services with PI = F301, F302,... , F316.

When the current ensemble contains service F211 then this is a reference service and is put first in the Id list. As there are in total 32 services to be put in the Id list, the list has to be split into four parts, carried in four different FIGs 0/6:

FIG index	OE	SIV	IdlQ	Id list
a	0	0 (start list)	01 (RDS PI)	F211 (reference), F301, ..., F311
b	0	1 (cont. list)	00 (DAB SId)	F212, F213, ..., F224
c	0	1 (cont. list)	00 (DAB SId)	F225, F226, F227
d	0	1 (cont. list)	01 (RDS PI)	F312, ..., F316

If none of the services belongs to the current ensemble, the OE flag is set to "1". In this case no reference identifier is defined so that the order of identifiers within the Id list is arbitrary.

Example: Same assumptions as above except that none of the services in the linkage set belong to the current ensemble:

FIG index	OE	SIV	IdlQ	Id list
a	1	0 (start list)	01 (RDS PI)	F301, F302, ..., F312
b	1	1 (cont. list)	00 (DAB SId)	F211, F212, ..., F223
c	1	1	01 (RDS PI)	F313, F314, F315, F316
d	1	1	00 (DAB SId)	F224, F225, F226, F227

2.5.5.3   Transmitter Identification Information Database

The Transmitter Identification Information (TII) database feature helps the receiver to determine its geographical location. TII codes can be inserted into the synchronisation channel of the transmitted DAB signal. The "pattern" and "comb" of the TII signals correspond to the MainId (Main identifier) and SubId (Sub-identifier) of the transmitter.

The TII database feature provides the cross-reference between these transmitter identifiers and the geographic location (in terms of coarse and fine longitude and latitude) of the transmitters. This can be useful in cases where a DAB receiver needs to make a decision to retune to another frequency and the decision criterion is marginal using the TII list alone. The TII database feature provides a much better specified area definition than is possible with only the geographical description in the region definition. If signals can be received from three or more transmitters, the receiver can use this information to perform a triangulation and pinpoint its position (see section 7.3.7).

The TII database feature is encoded in FIG type 0 extension 22 (FIG 0/22). The M/S flag allows differentiation between signalling the reference location of the MainId (M/S flag = 0) and signalling the location of transmitters relative to that reference location (M/S flag = 1). The location reference does not necessarily need to coincide with a transmitter location, but if it does then the latitude offset and longitude offset are set to "0" for that transmitter.

For terrestrial frequency planning it may be necessary to delay somewhat the various transmitter signals constituting an SFN in order to optimise the combined signal at the receiver. The TII database feature therefore comprises a TD (Time Delay) field, which signals the constant time delay (in microseconds) of the transmitter signal in the range 0 to 2047 µs.

In the receiver the latitude of the received transmitter is calculated by converting the two's complement value given in the combined Latitude coarse and fine fields (20 bits) to its decimal equivalent and then multiplying this value by $90°/2^{19}$. The longitude of the transmitter is calculated in a similar way but the decimal equivalent has to multiplied by $180°/2^{19}$. The resolution of latitude is 0.6'' which is about 19 metres, while the resolution of longitude is 1.2'' which corresponds to about 38 metres at the equator (or 19 metres at 60° latitude).

When M/S flag = 1 the location of the transmitters relative to the reference location is given in terms of the latitude and longitude offset of the transmitter from the reference associated with the same MainId. The maximum offset from the reference location that can be signalled is 5.625° of latitude and 11.25° of longitude.

The following example illustrates the case where the reference position corresponds to a transmitter located at 48°07'29'' North and 001°37'33'' West and which has in its null symbol a TII code with pattern number = "3" and comb number = "1", which means that Main Id = 0000011 and Sub Id = 00001. The latitude is +48°07'29'' with 07' = $(07/60)°$ = 0.11666° and 29'' = $(29/3600)°$ = 0.008055° so that +48°07'29'' = 48.1247222°.

In FIG 0/22, for M/S = "0", the latitude is coded as a 20-bit number comprising a 16-bit latitude coarse field and a 4-bit latitude fine field. A two's complement number has to be calculated such that when the decimal equivalent "L" of this number is multiplied by $90°/2^{19}$ the result is +48.1247222°. $L \cdot 90°/2^{19}$ = +48.1247222° so that L = +280346.8262 = 0100 0100 0111 0001 1010 or 0x4471A.

Table 2.15   Example of calculation of the latitude/longitude coarse and fine fields in FIG 0/22

	Latitude Coarse	Longitude Coarse
Geographical co-ordinates	48°07'29'' (North)	001°37'33'' (West)
In decimal degrees	+48.1247222°	−1.6258333°
Formula	$L \cdot (90°/2^{19})$ = +48.124722	$L \cdot (180°/2^{19})$= −1.625833
Decimal value of L	+280346	−4735
Binary value of L	0100 0100 0111 0001 1010	0000 0001 0010 0111 1111
One's complement of L		1111 1110 1101 1000 0000
Two's complement of L		1111 1110 1101 1000 0001
Value to be put in coarse field	0100 0100 0111 0001	1111 1110 1101 1000
	(0x4471)	(0xFED8)
Value to be put in fine field	1010	0001

Since L is a positive number the two's complement of L corresponds to the binary value. The 16 MSBs of this number are put in the latitude coarse field and the 4 LSBs are put in the latitude fine field. The values for the longitude coarse and fine fields can be derived in a similar way. Since the longitude is West of Greenwich, value L is negative. Table 2.15 shows how the values to be put in the longitude coarse and fine fields are derived from the two's complement number.

This results in the following coding of FIG 0/22 for M/S = 0 (Main Identifier):

M/S	MainId	Latitude coarse	Longitude coarse	Latitude fine	Longitude fine
0	0000011 (3)	0100 0100 0111 0001 (0x4471)	1111 1110 1101 1000 (0xFED8)	1000 (8)	0101 (5)

Since transmitter 3/1 is the reference position the latitude offset and longitude offset are equal to zero so that FIG 0/22 for M/S = 1 for transmitter 3/1 would look like:

M/S	MainId	SubId	Time delay	Latitude offset	Longitude offset
1	0000011 (3)	00001 (1)	00000000000 (0 μs)	0000 0000 0000 0000 (0x0000)	0000 0000 0000 0000 (0x0000)

Suppose that around this reference position in the SFN there are three transmitters (3/2, 3/3, 3/4) with MainId = 0000011 (3) and with SubId = 00010 (2), 00011 (3) and 00100 (4). Suppose that the latitude offset for transmitter 3/2 is 00°13'14'' North and the longitude offset is –000°14'15'' West (see Table 2.16). The time delay for this transmitter is 14 μs.

Table 2.16   Example of calculation of the latitude/longitude offset fields in FIG 0/22

	Latitude Offset	Longitude Offset
Geographical co-ordinates	00°13'14'' North	000°14'15'' West
Decimal degrees	+0.220555°	–0.2375°
Formula	$L \cdot (90°/2^{19}) = +0.220555°$	$L \cdot (180°/2^{19}) = -0.2375°$
Decimal value of L	+1285	–692
Binary value of [L]	0000 0101 0000 0101	0000 0010 1011 0100
One's complement of L		1111 1101 0100 1011
Two's complement of L		1111 1101 0100 1100
Hexadecimal value of L	0x0505	0xFD4C

FIG 0/22 for M/S = 1 (Sub Identifier) for transmitter 3/2 could look like:

M/S	MainId (7 bits)	SubId (5 bits)	Time delay (11 bits)	Latitude offset (16 bits)	Longitude offset (16 bits)
1	0000011 (3dec)	00010 (2dec)	00000001110 (14 μs)	0000 0101 0000 0101 (0x0501)	1111 1101 0100 1100 (0xFD4C)

The co-ordinates of transmitter "3/2" are +48°7'29'' + 00°13'14'' = +48°20'43'' latitude North and -001°37'33'' - 000°14'15'' = -001°51'48'' longitude West.

2.5.5.4 Region Definition and Region Label

Geographical regions within an ensemble coverage area can be identified by means of the region definition feature of FIG type 0 extension 11 (FIG 0/11) and the region label feature of FIG type 1 extension 3 (FIG 1/3). The region definition feature, which is based on lists of transmitter identification codes, allows a receiver to determine roughly where it is located. This knowledge allows a receiver to filter out information which is relevant only to that region. For example, a much shorter list of alternative frequencies can be checked for service following and alarm announcements can be confined to the region concerned.

The region label feature of FIG 1/3 provides a textual description of a region (e.g. Bretagne, Oberbayern, Kent, etc.). Region labels help a listener to select a region in conjunction, for example, with the announcement feature. Traffic information concerning the destination and the route can be requested as well as that for the starting point of a journey. The user can select, say, "Kent" from a region menu if he or she is interested in getting information from this region or for filtering (traffic or weather) announcements.

A region is uniquely identified by means of an 11-bit RegionId, which is composed of two parts: a 5-bit upper part and a 6-bit lower part. This allows up to 2047 regions to be identified within an ensemble coverage area. The region label uses only the 6-bit lower part (the 5-bit upper part is set to all zeros) so that up to 64 regions per ensemble can be labelled.

Description of a region

With FIG 0/11 there are two ways to describe a region:

1. By means of a list of transmitter identifiers (TII list). In this case a RegionId is associated to a group of transmitters with one common MainId and different SubIds. By means of this list the receiver can determine in which region it is located. This is useful for receiver processing in the case of frequency information to select a suitable alternative frequency.

   Example:

Region Id	Main Id	SubId1	SubId2	SubId3	SubId4	SubId5
00000 10110 (0x002E) Greater London	0001011 (0x0B)	00001 Crystal Palace	00010 Alexandra Palace	00011 Guildford	00100 Bluebell Hill	00101 Reigate

2. By means of a spherical rectangle. In this case a RegionId is associated to a geographical area in terms of a spherical rectangle from which the co-ordinates of the South-west corner are given as latitude/longitude coarse coded as a 16-bit two's complement number. The size of the spherical rectangle is defined by the extent of latitude/longitude coded as a 12-bit unsigned binary number. Region identification based on the spherical rectangle allows navigation systems based on GPS to be used for automatic receiver location.

Example:

RegionId	Latitude coarse	Longitude coarse	Extent of latitude	Extent of longitude
00000000001 (0x001)	1101001000111010 (0xD23A) −32°11'12''(South)	1111100111010010 (0xF9D2) −008°41'26''(West)	000111111111 (0x1FF) (01°24'12'') ± 156 km	000010100000 (0x0A0) (000°52'44'') ± 100 km at equator

Importance of the G/E flag

The G/E flag in FIG 0/11 determines the validity area of the RegionId, that is whether the RegionId applies to the ensemble (E) coverage area only (e.g. the Bretagne ensemble), or to a "global" (G) coverage area, that is the area defined by the country Id and ensemble ECC (e.g. the whole of France).

Each ensemble provider in a country may use none, some or all of the globally defined RegionIds that it finds appropriate for its purposes. All other RegionIds may be defined for use ensemble-wide, even if they were globally defined through the above-mentioned agreement but not used as global ones in this ensemble.

Suppose that in France there is an agreement to use code 00000001110 (0x000E or 14dec) as a globally defined RegionId to denote "Bretagne". An ensemble provider in Paris may use this code in its Paris ensemble to denote "Versailles" on condition that the provider puts in FIG 0/11 the G/E flag = 0 (E) for RegionId 00000001110 and refrains from using it as the global code for "Bretagne".

For announcements in other ensembles or in FM the selected region allows the receiver to filter out only those announcements relevant to that region. However, RDS or OE services may not be available over the whole DAB coverage area. The receiver must therefore check as a background task, using the FI feature, which RDS or OE services are currently available at the receiver's location in order to prevent the receiver from retuning to an RDS or OE service and finding that it cannot be received. Regions must always be in terms of those defined for the tuned ensemble.

3

Audio Services and Applications

THOMAS BOLTZE, WOLFGANG HOEG and GERHARD STOLL

3.1   General

The DAB system is designed to provide radio services and additional data services. This chapter focuses on the main audio services as the central application. The audio services use MPEG Audio Layer II to provide mono, stereo and multichannel programmes to the listener. Although the first broadcast applications and receivers will only supply stereo programmes, the migration to multichannel services is nevertheless already included in the system.

Data services will become more and more important and one can currently see many new ideas on their use being developed and tested. Most advanced is the Internet, but it is easy to envisage that the data capacity of the DAB system can be used for very similar services. This chapter focuses on the main service of DAB – the audio broadcasting service – and the standard that is being used. We intend to give the reader an understanding of the features of the audio system and the possible pitfalls without going into too many details that the user (content provider, broadcaster and listener) has no influence on.

With the introduction of the Compact Disc and its 16-bit PCM format, digital audio became popular, although its bit rate of 706 kbit/s per monophonic channel is rather high. In audio production resolutions of up to 24-bit PCM are in use. Lower bit-rates are mandatory if audio signals are to be transmitted over channels of limited capacity or are to be stored in storage media of limited capacity. Earlier proposals to reduce the PCM rates have followed those for speech coding. However, differences between audio and speech

Digital Audio Broadcasting: Principles and Applications, edited by W. Hoeg and T. Lauterbach
©2001 John Wiley & Sons, Ltd.

signals are manifold since audio coding implies higher values of sampling rate, amplitude resolution and dynamic range, larger variations in power density spectra, differences in human perception, and higher listener expectations of quality. Unlike speech, we also have to deal with stereo and multichannel audio signal presentations. New coding techniques for high-quality audio signals use the properties of human sound perception by exploiting the spectral and temporal masking effects of the ear. The quality of the reproduced sound must be as good as that obtained by 16-bit PCM with 44.1 or 48 kHz sampling rate. If for a minimum bit rate and reasonable complexity of the codec no perceptible difference between the original sound and the reproduction of the decoded audio signal exists, the optimum has been achieved. Such a source coding system, recently standardised by ISO/IEC as MPEG-1 Audio (IS 11172) and MPEG-2 Audio (IS 13838), allows a bit-rate reduction from 768 kbit/s down to about 100 kbit/s per monophonic channel, while preserving the subjective quality of the digital studio signal for any critical signal. This high gain in coding is possible because the noise is adapted to the masking thresholds and only those details of the signal are transmitted which will be perceived by the listener.

3.2   Audio Coding

3.2.1   Basic Principles

Two mechanisms can be used to reduce the bit rate of audio signals. One mechanism is determined mainly by removing the redundancy of the audio signal using statistical correlation. Additionally, this new generation of coding schemes reduces the irrelevancy of

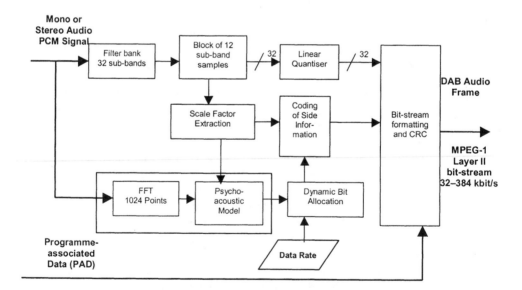

Figure 3.1   Block diagram of the ISO/IEC MPEG-1 Audio (IS 11172) Layer II encoder, as used in the EU147 DAB system

the audio signal by considering psycho-acoustical phenomena, like spectral and temporal masking (Zwicker, 1967). Only with both of these techniques, making use of the statistical correlation and the masking effects of the human ear, could a significant reduction of the bit rate down to 200 kbit/s per stereophonic signal and below be obtained.

The encoder block diagram of MPEG Audio Layer II, called "MUSICAM" (Dehéry, 1991) during its time of development and before standardisation in ISO/IEC, is shown in Figure 3.1. Consequently, Layer II is identical to the MUSICAM coding scheme, whereas Layer I has to be understood as a simplified version of the MUSICAM system. The basic structure of the coding technique which is more or less common to both, Layer I and Layer II, is characterised by the fact that MPEG Audio is based on perceptual audio coding.

One of the basic functions of the encoder is the mapping of the 20 kHz wide PCM input signal from the time domain into sub-sampled spectral components. For both layers a polyphase filterbank which consists of 32 equally spaced sub-bands is used to provide this functionality.

The output of a Fourier transform which is applied to the broadband PCM audio signal in parallel to the filter process, is used to calculate an estimate of the actual, time-dependent masked threshold. For this purpose, a psycho-acoustic model, based on rules from psycho-acoustics, is used as an additional function block in the encoder. This block simulates spectral and, to a certain extent, temporal masking too. The sub-band samples are quantised and coded with the intention to keep the noise, which is introduced by quantising, below the masked threshold. Layers I and II use a block companding technique with a scale factor consisting of 6 bits valid for a dynamic range of about 120 dB and a block length of 12 sub-band samples. With this kind of scaling technique, Layer I and Layer II can deal with a much higher dynamic range than CD or DAT, that is conventional 16-bit PCM.

In the case of stereo signals, the coding option "Joint Stereo" can be used as an additional feature. It exploits the redundancy and irrelevancy of typical stereophonic programme material, and can be used to increase the audio quality at low bit rates and/or reduce the bit rate for stereophonic signals (Waal, 1991). The increase of encoder complexity is small, and negligible additional decoder complexity is required. It is important to mention that Joint Stereo Coding does not enlarge the overall coding delay.

After encoding of the audio signal an assembly block is used to frame the MPEG Audio bit-stream which consists of consecutive audio frames.

The frame-based nature of MPEG Layer II leads to a delay through the encode/decode chain which is of the order of two frames (48 ms) when the processing and the tie for transmission are not taken into account. Typical delays through real transmission chains are of the order of 80 ms.

3.2.2   Masking of Audio Components

In order to reduce the bit rate of an audio signal whilst retaining the bandwidth, dynamic structure and harmonic balance, an audio codec has to reduce the quantisation step size of the signal. A CD signal is typically quantised with 16 bits per sample, which for a 0 dBFS sine wave results in a signal-to-quantisation-noise ratio of 96 dB. Reducing the quantisation step size will increase the quantisation noise relative to the audio signal. If the human auditory system were susceptible to this increase under all conditions, we could not reduce the bit rate.

Fortunately, the ear is a very imperfect organ. When we are exposed to a tone at one frequency, this excites the inner ear and the neurones in the cochlea in such a way that signals close (in frequency) to this signal may be inaudible, if their level is sufficiently low. This effect, which is called masking, decreases with the distance of the so-called masker and the masked signal.

A second imperfection is the fact that we do not hear audio signals below a certain level. This forms an absolute hearing threshold.

Both effects vary for human beings and have been measured in the past. They form the basis of what is called perceptual coding. The way to exploit these effects in a codec is to split the signal into small frequency regions and increase the quantisation noise in each frequency region such that it is as large as possible while still being masked by the signals in that frequency region.

The effect of masking of a single sine tone and its combination with the absolute hearing threshold is depicted in Figure 3.2. Two sine tones, which are masked, are shown in the figure as well.

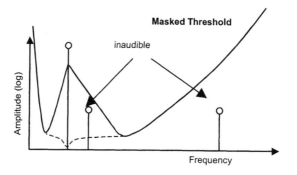

Figure 3.2   Masking of individual sine tone and absolute hearing threshold

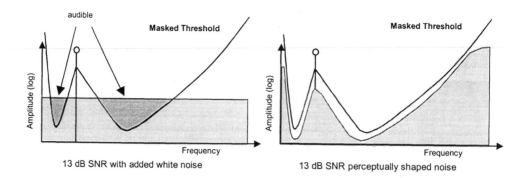

Figure 3.3   Added quantisation noise, without (left) and with (right) shaping

In Figure 3.3, the same amount of noise has been added to a sine tone. In the left part of the figure the noise has equal amplitude across all frequencies and is therefore audible. In the right half of the figure the noise has the same energy but has been shaped according to the masking curve. In this case the noise is inaudible. In the MPEG Audio codec the shaping of the noise is achieved by splitting the signal into 32 sub-bands before the reduction of the quantisation step size.

The masking properties have been measured for single sine tones and for narrow-band noise signals (Zwicker, 1967). The amount of masking depends on the nature of a signal and furthermore on the perceived tonality of a signal. The more tonal a signal is, the stronger the masking. It is therefore necessary to measure the tonality of a component to decide its masking properties. A ready assumption that is made in perceptual coding is that masking also occurs for complex signals, such as music and speech.

3.2.3   Psycho-acoustic Model

The psycho-acoustic model calculates the minimum masked threshold, which is necessary to determine the just-noticeable noise level for each band in the filterbank. The minimum masked threshold is the sum of all individual masking curves as described in the previous section. The difference between the maximum signal level and the minimum masked threshold is used in the bit or noise allocation to determine the actual quantiser level in each sub-band for each block. Two basic psycho-acoustic models are given in the informative part of the MPEG-1 standard (IS 11172). While they can both be applied to any layer of the MPEG Audio algorithm, in practice most encoders use model 1 for Layers I and II. In both psycho-acoustic models, the final output of the model is a signal-to-mask ratio for each sub-band of Layer II. A psycho-acoustic model is necessary only in the encoder. This allows decoders of significantly less complexity. It is therefore possible to improve the performance of the encoder even later, relating the ratio of bit rate and subjective quality. For some applications, which do not demand a very low bit rate, it is even possible to use a very simple encoder without any psycho-acoustic model.

The fundamental basis for calculating the masked threshold in the encoder is given by results of masked threshold measurements for narrow-band signals considering a tone masking noise and vice versa. Concerning the distance in frequency and the difference in sound pressure level, very limited and artificial masker/test-tone relations are described in the literature. The worst case results regarding the upper and lower slopes of the masking curves have been considered, assuming that the same masked thresholds can be used for both simple audio and complex audio situations.

The output of the FFT is used to determine the relevant tonal (i.e. sinusoidal) and non-tonal (i.e. noise) maskers of the actual audio signal. It is well known from psycho-acoustic research that the tonality of a masking component has an influence on the masked threshold. For this reason, it is worthwhile to discriminate between tonal and non-tonal components.

It is also known that the perception of the tonality of a component is time dependent. A pure tone is not perceived as tonal initially and builds up over time. This time dependency is part of more sophisticated psycho-acoustic models.

The individual masked thresholds for each masker above the absolute masked threshold are calculated depending on frequency position, loudness level and tonality. All the individual masked thresholds, including the absolute threshold, are added to the so-called global masked threshold. For each sub-band, the minimum value of this masking curve is determined. Finally, the difference between the maximum signal level calculated by both the scale factors and the power density spectrum of the FFT and the minimum masked threshold is calculated for each sub-band and each block.

The block length for Layer II is determined by 36 sub-band samples, corresponding to 1152 input audio PCM samples. This difference of maximum signal level and minimum masked threshold is called the signal-to-mask ratio (SMR) and is the relevant input function for the bit allocation.

3.2.4   The Filterbank

A high-frequency resolution, that is small sub-bands in the lower frequency region, whereas a lower resolution in the higher frequency region with wide sub-bands should be the basis for an adequate calculation of the masked thresholds in the frequency domain. This would lead to a tree structure of the filterbank. The polyphase filter network used for the sub-band filtering has a parallel structure, which does not provide sub-bands of different widths. Nevertheless, one major advantage of the filterbank is given by adapting the audio blocks optimally to the requirements of the temporal masking effects and inaudible pre-echoes. A second major advantage is the small delay and complexity. To compensate for the lack of accuracy of the spectrum analysis of the filterbank, a 1024-point FFT for Layer II is used in parallel with the process of filtering the audio signal into 32 sub-bands.

The prototype QMF filter is of order 511, optimised in terms of spectral resolution and rejection of sidelobes, which is better than 96 dB. This rejection is necessary for a sufficient cancellation of aliasing distortions. This filterbank provides a reasonable trade-off between temporal behaviour on one side and spectral accuracy on the other side. A time/frequency mapping providing a high number of sub-bands facilitates the bit rate reduction, due to the fact that the human ear perceives the audio information in the spectral domain with a resolution corresponding to the critical bands of the ear, or even lower. These critical bands have a width of about 100 Hz in the low-frequency region, that is below 500 Hz, and a width of about 20% of the centre frequency at higher frequencies.

The requirement of having a good spectral resolution is unfortunately contradictory to the necessity of keeping the transient impulse response, the so-called pre- and post-echo, within certain limits in terms of temporal position and amplitude compared to the attack of a percussive sound. Knowledge of the temporal masking behaviour (Fastl, 1977) gives an indication of the necessary temporal position and amplitude of the pre-echo generated by a time/frequency mapping in such a way that this pre-echo, which normally is much more critical compared to the post-echo, is masked by the original attack. In association with the dual synthesis filterbank located in the decoder, this filter technique provides a global transfer function optimised in terms of perfect impulse response perception.

3.2.5   Determination of Scale Factors

The sub-band samples are represented by a combination of scale factor and actual sample value. The scale factor is a coarse representation of the amplitude of either one individual sample or a group of samples. This approach allows for a higher reduction of the bit rate than to code each individual sample.

The calculation of the scale factor for each sub-band is performed for a block of 12 sub-band samples. The maximum of the absolute value of these 12 samples is determined and quantised with a word length of 6 bits, covering an overall dynamic range of 120 dB per sub-band with a resolution of 2 dB per scale factor class. In Layer I, a scale factor is transmitted for each block and each sub-band, which has no zero-bit allocation.

Layer II uses additional coding to reduce the transmission rate for the scale factors. Owing to the fact that in Layer II a frame corresponds to 36 sub-band samples, that is three times the length of a Layer I frame, three scale factors have to be transmitted in principle. To reduce the bit rate for the scale factors, a coding strategy which exploits the temporal masking effects of the ear has been studied. Three successive scale factors of each sub-band of one frame are considered together and classified into certain scale factor patterns. Depending on the pattern, one, two or three scale factors are transmitted together with additional scale factor select information consisting of 2 bits per sub-band. If there are only small deviations from one scale factor to the next, only the bigger one has to be transmitted. This occurs relatively often for stationary tonal sounds. If attacks of percussive sounds have to be coded, two or all three scale factors have to be transmitted, depending on the rising and falling edge of the attack.

This additional coding technique allows on average a factor of 2 in the reduction of the bit rate for the scale factors compared with Layer I.

3.2.6   Bit Allocation and Encoding of Bit Allocation Information

Before the adjustment to a fixed bit rate, the number of bits that are available for coding the samples must be determined. This number depends on the number of bits required for scale factors, scale factor select information, bit allocation information, and ancillary data.

The bit allocation procedure is determined by minimising the total noise-to-mask ratio over every sub-band and the whole frame. This procedure is an iterative process where, in each iteration step, the number of quantising levels of the sub-band that has the greatest benefit is increased with the constraint that the number of bits used does not exceed the number of bits available for that frame. Layer II uses only 4 bits for the coding of the bit allocation information for the lowest, and only 2 bits for the highest, sub-bands per audio frame.

3.2.7   Quantisation and Encoding of Sub-band Samples

First, each of the 12 sub-band samples of one block is normalised by dividing its value by the scale factor. The result is quantised according to the number of bits spent by the bit allocation block. Only odd numbers of quantisation levels are possible, allowing an exact representation of a digital zero. Layer I uses 14 different quantisation classes, containing $2n - 1$ steps, with $2 \leq n \leq 15$ different quantisation levels. This is the same for all sub-bands. Additionally, no quantisation at all can be used, if no bits are allocated to a sub-band.

In Layer II, the number of different quantisation levels depends on the sub-band number, but the range of the quantisation levels always covers a range of 3 to 65535 with the additional possibility of no quantisation at all. Samples of sub-bands in the low-frequency region can be quantised with 15, in the mid frequency range with seven, and in the high-frequency range only with three different quantisation levels. The classes may contain 3, 5, 7, 9, 15, 63, ..., 65535 quantisation levels. Since 3, 5 and 9 quantisation levels do not allow an efficient use of a codeword, consisting only of 2, 3 or 4 bits, three successive sub-band samples are grouped together into a "granule". Then the granule is coded with one codeword. The coding gain by using the grouping is up to 37.5%. Since many sub-bands, especially in the high-frequency region, are typically quantised with only 3, 5, 7 and 9 quantisation levels, the reduction factor of the length of the codewords is considerable.

3.2.8   Layer II Bit-stream Structure

The bit-stream of Layer II was constructed in such a way that a decoder of both low complexity and low decoding delay can be used, and that the encoded audio signal contains many entry points with short and constant time intervals. The encoded digital representation of an efficient coding algorithm specially suited for storage application must allow multiple entry points in the encoded data stream to record, play and edit short audio sequences and to define the editing positions precisely. For broadcasting this is important to allow fast switching between different audio programmes.

To enable a simple implementation of the decoder, the frame between those entry points must contain all the information that is necessary for decoding the bit-stream. Owing to the different applications, such a frame has to carry in addition all the information necessary for allowing a large coding range with many different parameters. In broadcasting frequent entry points in the bit-stream are needed to allow for an easy block concealment of consecutive erroneous samples impaired by burst errors.

The format of the encoded audio bit-stream for Layer II is shown in Figure 3.4. Short, autonomous audio frames corresponding to 1152 PCM samples characterise the structure of the bit-stream. Each audio frame starts with a header, followed by the bit allocation

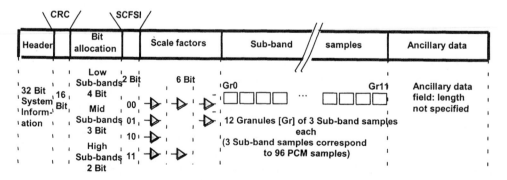

Figure 3.4   Audio frame structure of MPEG-1 Layer II encoded bit stream

information, scale factor and the quantised and coded sub-band samples. At the end of each audio frame is the so-called ancillary data field of variable length that can be specified for certain applications. Each frame can be accessed and decoded on its own. With 48 kHz sampling frequency, the frame duration is 24 ms for Layer II.

3.2.9   Layer II Audio Decoding

The block diagram of the decoder is shown in Figure 3.5. First of all, the header information, CRC check, the side information (i.e. the bit allocation information with scale factors), and 12 successive samples of each sub-band signal are extracted from the ISO/MPEG/AUDIO Layer II bit stream.

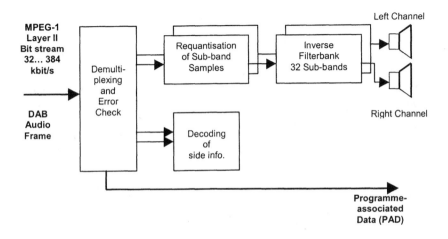

Figure 3.5   Block diagram of the MPEG-1 Audio Layer II decoder

The reconstruction process to obtain PCM audio again is characterised by filling up the data format of the sub-band samples regarding the scale factor and bit allocation for each sub-band and frame. The synthesis filterbank reconstructs the complete broadband audio signal with a bandwidth of up to 24 kHz. The decoding process needs significantly less computational power than the encoding process. The relation for Layer II is about 1/3. Because of the low computational power needed and the straightforward structure of the algorithm, Layer II can be easily implemented in special VLSI chips. Since 1993, stereo and multichannel decoder chips have consequently been available from many manufacturers.

3.3   Characteristics and Features of MPEG-1 Layer II for DAB

The EU147 DAB source coding system permits a digital audio broadcast receiver to use a standard MPEG-1 and MPEG-2 Layer II decoder. Besides the emission of the digital audio broadcast signal, the MPEG Audio Layer II coding technique and its encoded audio bit-stream can be used in a number of other applications, including contribution between broadcast studios, primary and secondary distribution, and news/sports reporting links. These different applications require a flexible coding scheme offering a wide range of parameters, in particular concerning the bit rates, audio modes (i.e. mono, stereo and multichannel representation), protection level of the coded bit-stream and the possibility to carry Programme-associated Data (PAD), which enable completely new applications.

More detailed information can be found in (Stoll, 1995) or (AES, 1996).

3.3.1   Audio Modes

The audio coding system, used in Eureka 147 DAB, supports the following modes:

- Mono (one-channel) mode.
- Stereo (two-channel) mode.
- Dual-channel mode. In this mode, the two audio channels can be either bilingual, or two mono channels, but with only one header. At the decoder a choice is made on which of the two programmes should be decoded.
- Joint Stereo mode. In the Joint Stereo mode, the encoder exploits the redundancy and irrelevancy of stereo signals for further data reduction. The method used for Joint Stereo in Layer II is "intensity stereo coding". This technique still preserves the spectral envelope of the left and right channel of a stereo signal, but transmits only the sum signal of the sub-band samples of the left and right channel in the high audio frequency region (Chambers, 1992).
- Low sampling frequency coding with f_S = 24 kHz.
- Provisions are made in the Eureka 147 DAB standard (EN 300401) for the inclusion of MPEG-2 Audio Layer II 5.1 backwards-compatible Surround Sound coding.

3.3.2   Sampling Rate and Input Resolution

The audio coding algorithm of DAB allows two sampling rates: 48 kHz and 24 kHz. The higher sampling rate can be chosen to have a full audio bandwidth of 20 kHz for the transmitted signal and to allow for a direct broadcasting of studio signals without the need for sampling-rate conversion. The audio quality of a PCM signal improves with increasing resolution of the input signal. Thus, the MPEG Audio Layer II standard can handle a resolution of the input signal up to 22 bits/sample.

The lower bit rate can be chosen to deliver a high quality, in particular for speech signals at very low bit rates, that is at bit rates ≤64 kbit/s per channel. However, this does not mean that a new sampling frequency will be introduced outside of the DAB system. Instead a downsampling filter from 48 kHz to 24 kHz in the audio encoder and an upsampling filter from 24 kHz to 48 kHz in the decoder are used.

Note: In contrast to other broadcasting systems (e. g. ADR) a sampling frequency of 44,1 kHz is not applicable, which means a source taken from a CD has to be sample-rate converted.

3.3.3   DAB Audio Frame Structure

The DAB audio frame is based on the MPEG Layer II audio frame which includes all the main and side information which is necessary for the decoder to produce an audio signal at its output again. Additionally, each DAB audio frame contains a number of bytes that may be used to carry the PAD, that is the information which is intimately related to the audio programme. This structure is depicted in Figure 3.6. The PAD field contains 2 bytes of Fixed PAD (F-PAD), and an optional extension called the extended PAD (X-PAD). Functions available for the PAD include Dynamic Range Control (DRC), music/speech indication, programme-related text and additional error protection. The PAD features are explained in section 3.4.

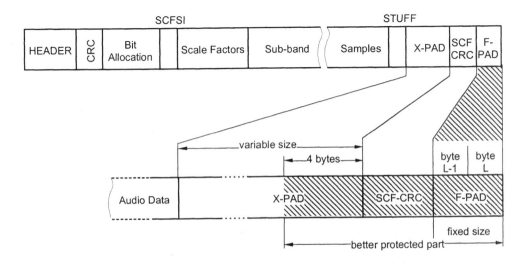

Figure 3.6   Structure of Fixed and Extended Programme-associated Data (F-PAD and X-PAD) and Scale Factor CRC (SCF-CRC) applied to the MSBs of the scale factors for either four or two sub-band groups, depending on the bit rate

3.3.4   Audio Bit Rates

Depending on the type of programme, the number of programmes within the DAB multiplex and the protection level, different bit rates in the range of 32 to 192 kbit/s in single channel mode can be selected at the audio encoder. The choice of the bit rate is not fully independent of the audio mode. Depending on the bit rate, the audio modes given in Table 3.1 can be used. Any combination of these options may be used within the DAB multiplex (Chambers, 1992). For the half-sampling-rate coding of 24 kHz, 14 different bit rates in the range of 8 to 160 kbit/s can be chosen, irrespective of the selected audio mode. For the multichannel extension, Table 3.1 is valid for the base bit-stream only. The details of the lower sample rates and the multichannel extension are explained in section 3.5.

The wide range of bit rates allows for applications that require a low bit rate and high audio quality; for example, if only one coding process has to be considered and cascading can be avoided. It also allows for applications where higher data rates up to about 180 kbit/s per channel could be desirable if either cascading or post-processing has to be taken into account. Experiments carried out by the ITU-R (AES, 1996) have shown that a coding process can be repeated nine times with MPEG-1 Layer II without any serious subjective degradation, if the bit rate is high enough, that is 180 kbit/s per channel. If the bit rate is only 120 kbit/s, however, no more than three coding processes should occur.

Table 3.1   Audio modes and bit rates for 48 kHz (left columns) and 24 kHz (right column) sampling rate

Sampling Rate	48 kHz				24 kHz
Mode	Single Ch.	Dual Ch.	Stereo	Int. Stereo	All
Bit Rate [kbit/s]					Bit Rate [kbit/s]
32	X				8
48	X				16
56	X				24
64	X	x	X	x	32
80	X				40
96	X	x	X	x	48
112	X	x	X	x	56
128	X	x	X	x	64
160	X	x	X	x	80
192	X	x	X	x	96
224		x	X	x	112
256		x	X	x	128
320		x	X	x	144
384		x	X	x	160

3.4   Programme-associated Data

The DAB system not only provides a very high audio quality for the listener, but also includes provision for many additional data services (BPN 007), (Riley, 1994). Most of these data services are independent of the audio programme, but some of them are closely related to the audio signal. The latter form is the so-called Programme-associated Data (PAD). A fixed and optionally a flexible number of bytes are provided to carry the PAD in the DAB audio bit-stream, see Figure 3.6. Additional capacity is provided elsewhere in the DAB multiplex (or "ensemble multiplex") to carry more independent data or additional information, such as text, still or moving pictures, etc., see Chapters 4 and 5.

The Fixed Programme-associated Data (F-PAD) are carrying amongst others:

- Dynamic Range Control (DRC) information, which may be used in the receiver to compress the dynamic range of the audio.
- Speech/music indication to allow for different audio processing of music and speech (e.g. independent volume control).

As these features are directly related to the audio programme signal, they are described in detail in the following. For more details on other programme-associated information carried within the F-PAD or X-PAD (Extended Programme-associated Data), see Chapters 2, 4 and 5.

In reproduction of audio broadcast programmes there will always be the need to apply dynamic range compression to some types of programme material because of the listeners' requirements. Also the satisfactory balancing of the loudness of different types of programme, particularly music and speech, depends principally upon the listeners' requirements. In conventional broadcasting it has never been possible to satisfy different individual habits of listening.

3.4.1   Functionality of Dynamic Range Control

The DAB DRC feature enables broadcasters to transmit programmes with a relatively wide dynamic range, accompanied by a DRC signal which the listener may use to effect unobtrusive compression of the programme dynamics, if required (Hoeg, 1994).

The dynamic range of an audio programme signal (sometimes termed the programme dynamic) is the range between the highest and the lowest useful programme signal level. The problems associated with programmes having a wide dynamic range, and with achieving a satisfactory loudness balance between different part of the radio programme (such as speech or music), are well known from experience with VHF/FM broadcasting (Müller, 1970). In many cases the source programme dynamic or the dynamic range commonly used by the broadcasters (approx. 40 dB) may be much larger than the usable dynamic range (the so-called reproduction dynamic) in noisy environments such as in a moving car. The reduction required in the dynamic range of the programme may be 10 to 30 dB (or more).

Taking into account the listening conditions and the requirements of the listener, these problems can only be solved at the receiver. For this to be feasible, additional information must be provided by the broadcaster concerning the gain adjustments, which may be needed to reduce the dynamic range. Therefore, at the broadcaster's premises a DRC signal is generated at the studio side, which describes the audio gain to be applied in the receiver, as a succession of values. This DRC signal is transmitted in a coded form (DRC data = DRCD) together with the audio signal.

It is a requirement of the DAB specification (EN 300401) that the audio signal is transmitted with its original programme dynamic, without any pre-compression. The DRC data are incorporated in the DAB bit-stream as PAD.

In the receiver, the regenerated DRC signal may be used optionally to control the audio gain in order to match the dynamic range of the received audio programme to the requirements of the listener, or to improve audibility in difficult conditions.

The user can choose any appropriate compression ratio between the anchors:

- *no compression*: the audio programme is reproduced with the dynamic range as delivered by the broadcaster, without any pre-compression before transmitting;
- *nominal compression*: the audio programme is reproduced with an appropriate dynamic compression as adjusted by the broadcaster (normally a compression ratio of about 1.3 is used);
- *maximum compression*: the audio programme can be reproduced with an extremely strong compression ratio (e.g. ≥ 2.0). This makes sense for poor reproduction conditions (e.g. in a car) and for special programme types only, because some influences on the sound quality cannot be excluded.

3.4.2   Functionality of Music/Speech (M/S) Control

It is well known from any kind of radio programme transmission that there may be very different requirements in the loudness balance between music and speech programmes, depending on the interests of the listener (Ilmonen, 1971). A music/speech (M/S) control signal, which is also transmitted in the DAB bit-stream, now enables the listener to balance the loudness of different types of programme according to taste.

There is the option to signal four states of the so-called M/S flag within the PAD:

00   programme content = Music

01   programme content = Speech

10   programme content = not signalled

11   reserved for future use (e.g. for a programme containing music and speech with artistically well-balanced levels, for instance a drama or musical).

This M/S flag information can be generated at the studio side automatically (e.g. derived from the corresponding channel fader at the mixing console). As an item of programme-related data, the information is transported within the PAD, and may be used at the receiver to control the output signal level (volume) in a predetermined manner, probably in a level range of about 0 to –15 dB, to enhance either the music, or the speech.

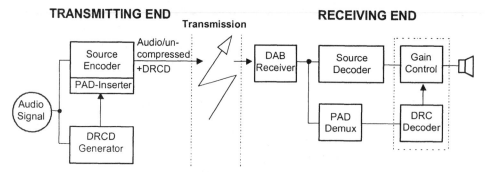

Figure 3.7   Block diagram of a DRC system for DAB

3.4.3   Transport Mechanism of DRC and M/S Data

Data travelling in the PAD channel are intimately related to the audio programme. They are assembled together with the coded audio in the same DAB frame, so the time relation established at the source is maintained throughout the complete DAB transmission chain. Audio and PAD must not be subject to different transmission delays.

Figure 3.7 shows a simplified block diagram of the DRC system for DAB, which is also valid in principle for transmission of the M/S control flag.

The DRC signal is generated by a characteristic which is linear over most of the dynamic range, with an appropriate compression ratio. An overall range of gain variation of about 15 dB is linearly coded in a 6-bit digital word, for each 24 ms frame of the broadcast audio signal.

The intention is to provide long-term compression of the dynamic range whilst preserving the impact of dramatic changes in the programme dynamics (e.g. that caused by a *crescendo* or *subito fortissimo*). Therefore the generating algorithms use specific sets of audio programme controlled time characteristics, which are not the subject of any specification.

The DAB audio programme signal itself is transmitted without any dynamic compression applied after the production/post-production stage. During the production of some types of programmes, compression may be applied for "artistic" reasons. However, programme signals compressed at this stage are treated in the same way as uncompressed signals by the DRC system.

Because the DRC data are transmitted discontinuously in time intervals of one DAB frame (24 ms) to avoid overshoots (i.e. brief periods in which the reproduced signal is excessively loud) in the output from the receiver, it is necessary to have a look-ahead time of one complete DAB frame in which to analyse the level of the signal. However, a delay is still needed in the signal path so that the incoming audio signal may be examined prior to the application of any changes in gain. This delay is for instance one frame length (24 ms), but may also be implemented with a significantly longer delay, up to several seconds.

3.4.5   The Receiving End

A common and simple receiver concept realises the necessary DRC functions. These functions do not depend upon the type of algorithm in the DRC signal generator used at the transmitting end. The same options for listening to the uncompressed programme, the programme with the nominal degree of compression and the programme with a greater degree of compression are always available.

The DRC data are demultiplexed from the PAD area, and a linear interpolation regenerates a continuous signal to control the gain of the audio channel, in order to match the dynamic range of the received programme to the individual listener's needs. The maximum gain change in the compression mode will be about 15 dB. If the error protection applied to the PAD fails, specifically defined limiting characteristics for the rate of gain change will protect the compressed audio signal against severe impairments.

The conventional way for controlling the dynamic range is to use a simple multiplication in the digital domain, or a VCA (or DCA) at the analogue end of the receiver to realise all necessary gain adjustments (DRC, manual or automatic volume control and other gain-related features such as the balancing of music and speech loudness).

3.4.6   Implementations

Solutions fulfilling the DAB requirements for DRC data have been developed by the BBC and Deutsche Telekom with Fa. Jünger Audio as well, see (Hoeg, 1994); there is also an existing solution from Swedish Broadcasting. The BBC system normally needs a significant look-ahead time, typically from 30 ms up to 3 s, but the Deutsche Telekom system typically needs only about 24 ms. In the meantime, there are also several configurations of DAB receiver equipment (see Chapter 8) which support the DRC and M/S functionality.

Besides the DRC system with control data being transmitted as described in the DAB standard (EN 300401), self-contained compressors for use in receivers which are operating *without* the need for transmitted control data have been proposed several times. One of these (Theile, 1993) is based on scale factor weighting in the MPEG-1 Layer II source decoder (the so-called MUSICAM-DRC system).

3.5   Half-sampling-rate and Multichannel Audio Coding

One of the basic features of the MPEG-2 Audio standard (IS 13818-3) is the backward compatibility to MPEG-1 (IS 11172) coded mono, stereo or dual-channel audio programmes. This means that an MPEG-1 audio decoder is able to properly decode the basic stereo information of a multichannel programme. The basic stereo information is kept in the left and right channels, which constitute an appropriate downmix of the audio information in all channels. This downmix is produced in the encoder automatically.

The backward compatibility to two-channel stereo is a strong requirement for many service providers which may provide high-quality digital surround sound in the future. There is already a wide range of MPEG-1 Layer I and Layer II decoder chips which support mono and stereo sound.

The use of multichannel audio is becoming adopted with the advent of digital terrestrial television (DVB-T, see section 1.5). With the backward compatibility of the MPEG multichannel audio coding standard, such a two-channel decoder will always deliver a correct stereo signal with all audio information from the MPEG-2 multichannel audio bit-stream.

MPEG-1 audio was extended as part of the MPEG-2 activity to lower sampling frequencies in order to improve the audio quality for mono and conventional stereo signals for bit rates at or below 64 kbit/s per channel, in particular for commentary applications. This goal has been achieved by reducing the sampling rate to 16, 22.05 or 24 kHz, providing a bandwidth up to 7.5, 10.5 or 11.5 kHz. The only difference compared with MPEG-1 is a change in the encoder and decoder tables of bit rates and bit allocation. The encoding and decoding principles of the MPEG-1 Audio layers are fully maintained.

3.5.1   Characteristics of the MPEG-2 Multichannel Audio Coding System

A generic digital multichannel sound system applicable to television and sound broadcasting and storage, as well as to other non-broadcasting applications, should meet several basic requirements and provide a number of technical/operational features. Owing to the fact that during the next few years the normal stereo representation will still play a dominant role for most of the consumer applications, two-channel compatibility is one of the basic requirements. Other important requirements are interoperability between different media, downward compatibility with sound formats consisting of a smaller number of audio channels and therefore providing a reduced surround sound performance. In order to allow applications as universal as possible, other aspects, like multilingual services, clean dialogue and dynamic range compression, are also important.

MPEG-2 audio allows for a wide range of bit rates from 32 kbit/s up to 1066 kbit/s. This wide range could be realised by splitting the MPEG-2 audio frame into two parts:

1. The primary bit-stream which carries the MPEG-1 compatible stereo information of maximal 384 kbit/s;
2. The extension bit-stream which carries either the whole or a part of the MPEG-2 specific information, that is the multichannel and multilingual information, which is not relevant to an MPEG-1 audio decoder.

The primary bit-stream realises a maximum of 448 kbit/s for Layer I and 384 kbit/s for Layer II. The extension bit-stream realises the surplus bit rate. If, in the case of Layer II, a total of 384 kbit/s is selected, the extension bit-stream can be omitted. The bit rate is not required to be fixed, because MPEG-2 allows for variable bit rate which could be of interest in ATM transmission or storage applications, such as DVD (Digital Video Disk).

Serious quality evaluations (BPN 019), (Wüstenhagen, 1998) have shown that a bit rate of about 512 to 640 kbit/s will be sufficient to provide a reasonable audio quality for multichannel audio transmission using MPEG-2 Layer II audio coding (non-coded transmission of five signals would need a net rate of $5 \times 768 = 3.840$ kbit/s).

3.5.2   3/2-stereo Presentation Performance

The five-channel system recommended by ITU-R (BS.775-1), SMPTE (RP200) and EBU (R96) is referred to as "3/2-stereo" (3 front/2 surround channels) and requires the handling of five channels in the studio, storage media, contribution, distribution, emission links, and in the home.

3.5.3   Low-frequency Enhancement Channel

According to (BS.775-1), the 3/2-stereo sound format should provide an optional Low-frequency Enhancement/Extension (LFE) channel in addition to the full range main channels. The LFE is capable of carrying signals in the frequency range 20 Hz to 120 Hz. The purpose of this channel is to enable listeners, who choose to, to extend the low-frequency content of the audio programme in terms of both low frequencies and their level. From the producer's perspective this may allow for smaller headroom settings in the main audio channels. In practice, the LFE channel is mainly used with film sound tracks. The mentioned 3/2 format is commonly called "5.1 format", where ".1" means the LFE channel.

3.5.4   Backward/forward Compatibility with MPEG-1

For several applications it is the intention to improve the existing 2/0-stereo sound system step by step by transmitting additional sound channels (centre, surround), without making use of simulcast operation. The multichannel sound decoder has to be backward/forward compatible with the existing sound format.

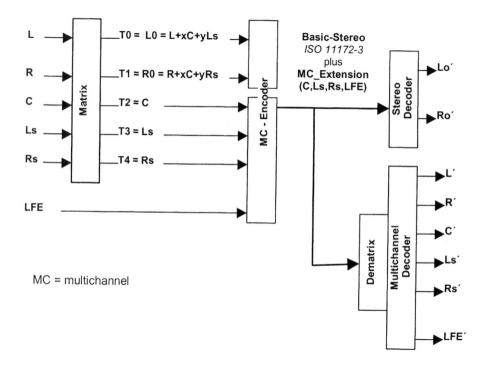

Figure 3.8   Backward compatibility of MPEG-2 Audio with MPEG-1

Backward compatibility means that the existing two-channel (low-price) decoder should properly decode the basic 2/0-stereo information from the multichannel bit-stream (see Figure 3.8). This implies the provision of compatibility matrices (Kate, 1992) using adequate downmix coefficients to create the compatible stereo signals L_o and R_o, shown in Figure 3.9. The inverse matrix to recover the five separate audio channels in the MPEG-2 decoder is also shown in the same figure.

The basic matrix equations used in the encoder to convert the five input signals L, R, C, LS and RS into the five transport channels T_0, T_1, T_2, T_3 and T_4 are:

$$T_0 = L_0 = (\alpha \times L) + \beta \times (\alpha \times C) + \gamma \times (\alpha \times LS)$$
$$T_1 = R_0 = (\alpha \times R) + \beta \times (\alpha \times C) + \gamma \times (\alpha \times RS)$$
$$T_2 = C^W = \alpha \times \beta \times C$$
$$T_3 = LS^W = \alpha \times \gamma \times LS$$
$$T_4 = RS^W = \alpha \times \gamma \times RS$$

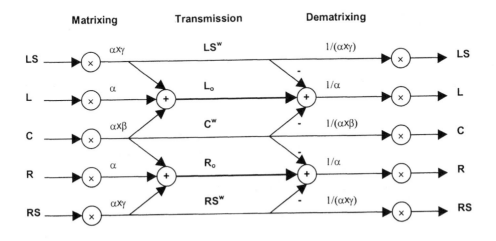

Figure 3.9 Compatibility matrix (encoder) to create the compatible basic stereo signal, and the inverse matrix (decoder) to re-establish the discrete five audio channels

In order to obtain maximal bit-rate reduction, T_2, T_3 and T_4 are also allowed to carry $(\alpha \times L)$ and/or $(\alpha \times R)$ instead of the listed $(\beta \times \alpha \times C)$, $(\gamma \times \alpha \times LS)$ and $(\gamma \times \alpha \times RS)$.

Four matrix procedures with different coefficients α, β and γ have been defined and can be chosen in the MPEG-2 multichannel encoder. Three of these procedures add the centre signal with 3 dB attenuation to the L and R signals. The surround signals LS and RS are added to the L, respectively R, signals with either 3 dB or 6 dB attenuation. The possibility of an overload of the compatible stereo signals L_o and R_o is avoided by the attenuation factor α which is used on the individual signals L, R, C, LS and RS prior to matrixing. One of these procedures provides compatibility with Dolby Surround® decoding. Being a two-channel format, compatibility can already be realised in MPEG-1. MPEG-2 allows such transmissions to be extended to a full, discrete five-channel size.

The fourth procedure means that no downmix is included in the bit-stream, which actually constitutes a "Non-backwards Compatible" (NBC) mode for the MPEG-2 multichannel codec. An MPEG-1 decoder will produce the L and R signals of the multichannel mix. In certain recording conditions this "matrix" will provide the optimal stereo mix.

Forward compatibility means that a future multichannel decoder should be able to decode properly the basic 2/0-stereo bit-stream.

The compatibility is realised by exploiting the ancillary data field of the ISO/IEC 11172-3 audio frame for the provision of additional channels (see Figure 3.10). The "variable length" of the ancillary data field gives the possibility to carry complete multichannel extension information. If for Layer II the bit rate for the multichannel audio signal exceeds 384 kbit/s, an extension part is added to the MPEG-1 compatible part.

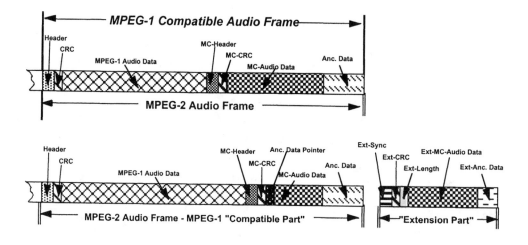

Figure 3.10 ISO/IEC 13818-3 (MPEG-2 Audio) Layer II multichannel audio frame consisting of the MPEG-1 compatible part and the extension part

However, all the information about the compatible stereo signal has to be kept in the MPEG-1 compatible part. In this case, the MPEG-2 audio frame consists of the MPEG-1 compatible and the (non-compatible) extension parts. This is shown in Figure 3.10.

In the first generation the DAB system (EN 300401) will not provide multichannel sound. Therefore the extension to digital surround sound has to be backward/forward compatible with an MPEG-1 Audio decoder.

3.5.5   Compatibility with Matrix Surround

Matrix surround systems, such as Dolby ProLogic®, have found wide acceptance in consumers' homes. Many movies and drama programmes are produced not only in discrete multichannel but also in matrix surround for delivery on video tape and analogue broadcast.

To broadcasters this legacy of matrix surround systems means that they should be able to continue to broadcast matrix surround material with the digital system. The audio coding system should be compatible with the matrix surround system.

The BBC conducted a very good study on the compatibility of MPEG Layer II and Dolby ProLogic® (then the dominant matrix surround system) and found that, provided some precautions are taken, the surround information is completely preserved (Meares, 1998).

3.5.6   Multilingual Extension and Associated Services

MPEG-2 Audio provides alternative sound channel configurations in the multichannel sound system, for example the application of the "second stereo programme" might be a bilingual 2/0-stereo programme or the transmission of an additional binaural signal. Other configurations might consist of one 3/2 surround sound plus accompanying services (e.g. clean dialogue for the hard-of-hearing, commentary for the visually impaired, multilingual commentary, etc.). For these services, either the multilingual extension or the ancillary data field, both provided by the MPEG-2 bit-stream, can be used.

A good case for providing a multilingual service in combination with surround sound is given when the spoken contribution is not part of the acoustic environment that is being portrayed. Surround sound sports effects can be combined with multiple language mono commentary channels. In contrast, surround sound with drama would require a new five-channel mix for each additional language.

An important issue is certainly the "final mix in the decoder"; that means the reproduction of one selected commentary/dialogue (e.g. via centre loudspeaker) together with the common music/effect stereo downmix (examples are documentary film, sport reportage). If backward compatibility is required, the basic signals have to contain the information of the primary commentary/dialogue signal, which has to be subtracted in the multichannel decoder when an alternative commentary/dialogue is selected.

In addition to these services, broadcasters should also be considering services for hearing-impaired consumers. In this case a clean dialogue channel (i.e. no sound effects) would be most advantageous. This could be transmitted at a low bit rate of about 48 kbit/s with the lower sampling frequency coding technique. This provides excellent speech quality at a bit rate of 64 kbit/s and even below, making very little demand on the available capacity of the transmission channel (BPN 021).

3.6   Quality of Service

The quality of service in digital transmission systems is typically much more constant than with analogue transmissions, such as FM or AM. The digital transmission system provides a very robust channel for the compressed audio data. Variations in the quality of the transmission channel, fading effects, shadowing and interference with other sources that affect the audio quality directly in an analogue system will have much less influence on the perceived quality in a digital system. The reason for this is that the digital modulation, together with the error correction, protect the audio data from corruption.

As long as the error correction can correct the influence of the channel, no change in the audio stream will occur, resulting in a constant quality over time. Only when the error correction fails will the audio be impaired. If this happens, the severity of the impairment depends on which part of the audio frame is corrupted.

Under the assumption that the error correction works, the quality of service in a digital system is mainly determined by the combination of bit rate, codec implementation and actual programme material. Another factor is the listener. Experience shows that what might be an inaudible artefact when heard first might become very obvious when the listener learns to identify it as an artefact. This is true for any kind of coding scheme and also holds in the field of video compression.

3.6.1   Quality Versus Bit Rate

The perceived audio quality depends on the bit rate in a non-linear fashion. Towards higher bit rates saturation around imperceptible differences with the original can be observed, while the quality suddenly drops towards lower bit rates. At this point the coding scheme simply cannot provide satisfactory results and a number of different artefacts (such as quantisation noise, bandwidth limitation and change of the stereo image) occur all at once.

The MPEG-1 standard (as it is used on DAB) uses a sample rate of 48 kHz. MPEG-2 has defined an extension to 24 kHz in addition. Both are allowed in DAB.

The motivation behind the extension to lower sample rates was to provide a better frequency resolution for signals which are band limited by nature. These are typically speech programmes, which constitute a significant part of broadcasts.

The limitation to a bandwidth of about 11 kHz results in a higher frequency resolution at frequencies where the ear is very sensitive to errors. This results in a crossover point between 48 kHz and 24 kHz, at which it is more attractive to use either 48 kHz or 24 kHz, depending on the actual bit rate and the programme material.

The data presented in Figure 3.11 and Figure 3.12 are taken from (EBU, 1999). It is a summary of several different listening tests, performed by several international organisations. The listening tests were executed according to the ITU rules and the scale

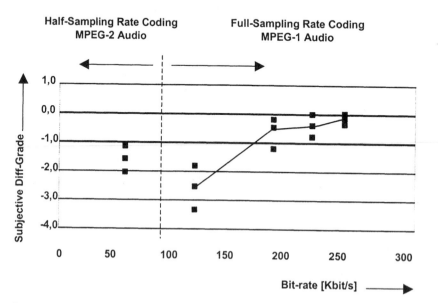

Figure 3.11   MPEG Layer II, stereo

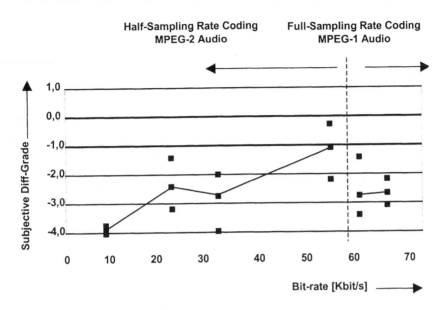

Figure 3.12   MPEG Layer II, mono

used is that of absolute impairment when compared to the original. The three points for each bit rate give the mean score for the least critical item and the most critical item on the upper and lower end and the mean score over all items in the middle.

We mentioned earlier that the audio quality for a certain bit rate and setting depends on the actual codec implementation, because the psycho-acoustic model and the bit allocation strategy are not part of the standard.

Experience with developing encoders shows that the differences in audio quality can be fairly significant. The results given above can only be an indication of the general trend, but should not be used as the only rule to set all the coding parameters in the daily practice. A further reason for this lies in the nature of listening tests: in every well-organised listening test, most of the time is spent on finding the most critical material for all encoders involved, whilst maintaining a balance between different types of material and the different encoders. This leads to results for the worst case of known material, but not for typical material.

Further information on the relationship between quality and bit rate can be found in (AES, 1996).

3.7   Error Protection and Concealment

Only very limited error protection for the audio bit-stream is provided in the MPEG standard. Error protection has to take into account the characteristics of the source data and the transmission channel. The MPEG Audio standards have been written for a wide range of applications with very different transmission channels, ranging from nearly completely transparent (i.e. error-free transmission channels, like storage on computer hard disks) to very hostile transmissions paths, like mobile reception with DAB (EN 300401).

3.7.1   Error Protection

In the presence of only a very few errors, say with a bit error rate of about 10^{-5} to 10^{-6} and lower, the optional CRC check, provided in the ISO standard, will in general be an efficient tool to avoid severe impairments of the reconstructed audio signal. Errors in the most sensitive information, that is header information, bit allocation (BAL) and scale factor select information (SCFSI), can be detected. The odds of bit errors in this part of the audio frame are small. If, however, one single bit error occurs in these fields, the result will be the loss of a complete audio frame. In this case, the result of a single bit error is the same as if a complete audio frame is lost by a burst error or cell loss.

To protect a listener of ISO/MPEG Audio coded audio signals from annoying distortions due to bit errors, channel coding has on the one hand the task to correct as many bit errors as possible, and on the other hand to enable a detection of residual bit errors. In the Eureka 147 DAB system, the data representing each of the programme services being broadcast is subjected to energy dispersal scrambling, convolutional coding and time interleaving. The convolutional encoding process involves adding redundancy to the service data using a code with a constraint length of 7. In the case of an audio programme, stronger protection is given to some bits than others, following a pre-selected pattern known as the Unequal Error Protection (UEP) profile, shown in Figure 3.13. The average code rate, defined as the ratio between the number of source-encoded bits and the number of encoded bits after convolutional encoding, may take a value from 0.33 (the highest protection level) to 0.75 (the lowest protection level). Different average code rates can be applied to different

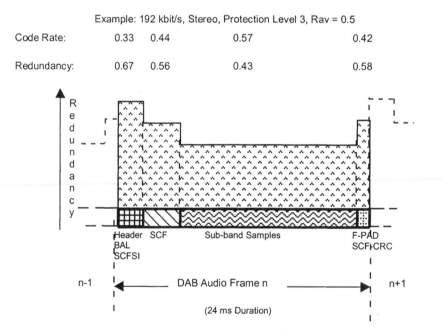

Figure 3.13   Unequal error protection scheme used in the DAB system for an ISO/IEC 11172-3 Layer II audio frame

audio sources. For example, the protection level of audio services carried by cable networks may be lower than that of services transmitted in radio frequency channels.

Even with the best error correction scheme residual bit errors cannot be completely avoided, and have to be detected, especially for information which is very sensitive to bit errors.

This includes the header and control information, which has been taken into account by the ISO CRC, but also for scale factors, which has been only partially considered by ISO/MPEG.

The optimisation criterion for protecting encoded audio signals against bit errors is not the minimisation of the bit error ratio as the most important issue, but to minimise the perception of audio signal distortions in the case of bit errors. The subjective annoyance of bit errors depends strongly on the kind of disturbed data for the ISO/MPEG Audio coded signal.

Whereas a bit error rate of 10^{-4} for sub-band samples results in barely audible degradation, any single bit error in the control information (header, BAL or SCFSI) causes a total frame error of 24 to 36 ms. The individual bit error ratios have to be balanced with respect to a certain degradation with the goal of minimising the overall distortion. Besides the economical aspect due to minimal additional redundancy, a UEP scheme allows for "graceful degradation" which may be very convenient for the listener in the case of an increase of the bit error rate.

In the case of the Eureka 147 DAB system, the following measures for residual error detection strategies can be used:

Frame CRC (First Cyclic Redundancy Check), specified by ISO/IEC (IS 11172). A 16-bit parity check word can be used for error detection of the main audio information within the encoded bit-stream, that is ISO header, BAL and SCFSI. If the CRC is violated a concealment based on non-reliable frames is requested.

Scale factor CRC (Second Cyclic Redundancy Check), specified by Eureka 147/ETSI (EN 300401). Up to four 8-bit parity check words can be used for error detection of single scale factors within sub-band groups. If the CRC is violated a concealment based on non-reliable scale factors is requested.

Reliability information. If conventional or punctured convolution codes are used for the error protection, additional reliability information could be derived by a Viterbi channel decoder. Reliability information derived by the channel decoder will give more information on the actually violated data and would support the adaptation of concealment to the error situation.

Failure characteristics of the DAB system have also been tested by subjective assessment, see for instance (Lever, 1997).

3.7.2   Concealment Measures

Error correction and error detection strategies can be optimised concerning both the encoded audio signal and the listener's perception. Although some measures for an error correction may have been applied, the channel coding technique has to provide provisions for the detection of residual bit errors, because every channel code can be overloaded. Several error concealment strategies in combination with low bit rate coding, like muting, substitution, repetition or estimation of the disturbed parts of the received encoded audio data, have already been proposed and described in (Dehéry, 1991). Those techniques improve the audio quality in case of transmission errors, but some of them, for example left/right-substitution, are not compatible with the audio bit-stream, and it cannot be expected that those techniques will be applied.

Applying error concealment techniques should result in the following improvements for the listener:

- Improved sound quality at the edge of the coverage area (single and burst errors)
- Improved sound quality during bad mobile reception conditions (single and burst errors)
- Improved sound quality during cell loss in ATM networks (burst errors)
- Performance and costs trade-offs between simple and cheap receivers and sophisticated and expensive receivers.

None of these benefits are available to the user in the old analogue system.

3.7.3   Frame-oriented Error Concealment

A complete frame or even some successive frames have to be processed with frame-oriented error concealment if the frame CRC indicates an error. Depending on the receiver's capabilities several concealment techniques can be applied.

Muting. The simplest method of frame-related error concealment is to mute frames which are indicated as non-reliable or even non-decodable. Compared to a very annoying reception the subjective annoyance caused by occasional muting of these frames might be assessed as lower.

Repetition. Instead of muting, non-decodable frames can be replaced by previously correct decoded frames. The audio quality of a repetition depends strongly on the signal itself; mostly it is much better compared to a simple muting. Contrary to muting, the repetition of frames needs memory of at least one frame (768 bytes at a bit rate of 256 kbit/s). A longer break of some frames can be bridged by either multiple repetition of one frame or, depending on the memory provided in a receiver, by a single repetition of the necessary number of previously correct decoded frames.

3.7.4   Error Concealment of Scale Factors

Scale factors represent the maximum absolute value of a sequence of samples in each sub-band. Transmission errors of scale factors do not lead to a distortion of one complete frame, but can be compared to a change produced by an equaliser in a way where one or more sub-bands are amplified or attenuated by the disturbed scale factor. If MSBs of scale factors are not decoded correctly, "beeping"-type impairments are produced and perceived as "very annoying".

The probability density function of differences of successive scale factors shows that for 90% of the time there are only small changes in the range of plus or minus one scale factor class. It is very difficult to detect such an error, which is limited to one sub-band and to an amplification or attenuation of about 2 dB.

Based on these statistical data of scale factors, a very simple error concealment can be applied. The disturbed scale factors are replaced by previously correct decoded scale factors. This can be considered as a concealment technique with no or only small impairments of quality.

3.7.5   *Error Concealment by Plausibility Checks*

The statistical properties of different information within one audio frame consisting of a DAB header, BAL, SCFSI, scale factors and sub-band data can be exploited for error detection.

An analysis of the successive scale factors of each sub-band from the point of view of statistical properties of audio signals can give additional and more detailed information on the reliability of the received scale factors. During an attack of the audio signal (e.g. a piano chord), scale factors may change greatly within a short time and may be followed by scale factor values which decrease slowly. During the decay of an audio signal continuously decreasing values of scale factors can be expected. This means that a high increase followed by a high decrease is not plausible. Thus, the scale factor has to be indicated as non-reliable and has to be concealed by, for example, repetition or interpolation. Compared to other possibilities of error protection and detection codes for the scale factors, this type of plausibility analysis does not need additional channel capacity and thus may lead to a similar evaluation of the error situation and performance.

In addition the received scale factor values can be compared to the range of values expected by the SCFSI. If the SCFSI indicates the transmission of three scale factors within one sub-band, a certain difference (depending on the encoder algorithm) between the successive scale factors can be assumed. A more detailed investigation of these interdependencies might lead to a basis for a predictor model comparing received audio data with predicted data and classifying the received audio data as more or less reliable.

3.8   A Typical DAB Ensemble

The DAB system is based on a multiplex that can assign different bit rates to each individual programme and that also allows dynamic reconfiguration. The latter could be used to split one high-quality programme into two of lower quality or vice versa.

The actual protection ratio and bit-rate for each individual programme depend on the codec used, the material for that programme and the intended audience. Although it would be desirable from a quality point of view always to use high bit rates, this is not economical. More programmes at acceptable quality are a better deal than fewer programmes at guaranteed CD quality.

Table 3.1 below provides possible values taken from (BPN 007). Table 3.2 lists the type of programme with recommended bit rates and channel configurations for certain quality levels. We can see a fairly wide range of bit rates for all different quality levels. The coding mode (stereo or Joint Stereo) is not included in this table, but experience shows that the use of Joint Stereo coding can reduce the required bit rate. Again the existing codec implementations differ significantly, and the actual audio quality therefore does so too.

Table 3.2   Sets of suitable parameters for different materials and quality levels of audio services

Programme Type	Format *	Quality Level	Sampling Rate kHz	Protection Level	Bit Rate kbit/s
Music/Speech	1/0	Broadcast quality	48	UEP_2 or 3	112 to 160
	2/0	Broadcast quality	48	UEP_2 or 3	128 to 224
	3/2	Broadcast quality	48	UEP_2 or 3	384 to 640
Speech only	1/0	Acceptable quality	24 or 48	UEP_3	64 to 112
Messages	1/0	Intelligible message	24 or 48	UEP_4	32 or 64
Data	Non			UEP_4	32 to 64

* Format codes:	1/0	mono
	2/0	two-channel stereo
	3/2	multichannel

An actual ensemble, as broadcast in Bavaria, Germany at the time of writing of this book, is summarised in Table 3.3. The bit rates reflect the development that has taken place over the years in the encoders used. They are slightly lower than in (BPN 007). Also joint coding is employed for all stereo programmes.

As more and more countries start their DAB services, more information will be come available. The figures taken from the Ensemble Bayern can be regarded as fairly typical and should be a good starting point for a broadcaster to plan a service. Another actual multiplex configuration is also described in Chapter 5, Table 5.1.

Table 3.3   Actual values for Ensemble "Bayern"

Sub-channel	Bit Rate kbit/s	Audio Mode	Protection Level	Service Component
2	160	Joint Stereo	UEP_3	Bayern 2 Radio
3	160	Joint Stereo	UEP_3	Bayern 3
4	192	Joint Stereo	UEP_3	Bayern 4 Klassik
5	96	Single Channel	UEP_3	Bayern 5 Aktuell
1	160	Joint Stereo	UEP_3	Bayern Mobil
7	192	Joint Stereo	UEP_3	Rock Antenne
6	160	Joint Stereo	UEP_3	Radio Galaxy
11			UEP_4	MobilData BDR
11			UEP_4	Update

4

Data Services and Applications

ROLAND PLANKENBÜHLER, BERNHARD FEITEN and
THOMAS LAUTERBACH

4.1   General

While DAB development started mainly focused on *audio* broadcast, data capabilities have been designed into the system right from the beginning. After successful presentations of the new digital audio broadcast system at the beginning of the 1990s, these data capabilities gained more and more interest and prototype data services were demonstrated by different project partners of Eureka 147 the related events (e.g. at NAB in Las Vegas, Spring 1992). In order to harmonise these prototype data services and to reach compatibility between the equipment and services, data interfaces, protocols and applications were defined to match the needs of broadcasters, service providers and end users. In the meantime it is generally accepted that the usage of some DAB channel capacity, especially for multimedia data services, is crucial both for user acceptance and for the financing of the networks of this new digital broadcast medium. All the necessary hooks are available to follow this route.

4.2   Multimedia Applications with MOT

4.2.1   The Multimedia Object Transfer Protocol

4.2.1.1   General Principles
The "Multimedia Object Transfer" (MOT) Protocol (EN 301234) is a transport protocol for the transmission of multimedia content in DAB data channels to various receiver types with multimedia capabilities. It is the common transport mechanism for transmitting information

Digital Audio Broadcasting: Principles and Applications, edited by W. Hoeg and T. Lauterbach
©2001 John Wiley & Sons, Ltd.

via different DAB data channels (PAD, packet mode) to unify access to multimedia content within the DAB system. MOT ensures interoperability between different data services and application types, different receiver device types and targets, as well as for equipment from different manufacturers. A basic support for multimedia presentations is also provided. For implementation details and hints see (TR 101497).

4.2.1.2   Structure of MOT

All information is conveyed from the information source to the destination as *objects* of finite length. The maximum object length which can be signalled is about 255 Mbytes. No constraints on the content which can be transmitted are applied. Currently data types like HTML, JPEG, MPEG audio/video and many more are defined (see (EN 301234) for the complete list), but this list can also be extended in the future if there is a demand for new data type/subtype entries. User applications may define specific methods to signal multimedia data types, for example the user application "Broadcast Web Site" uses "Mime Type" for signalling.

Each object consists of a *header* and a *body* (see Figure 4.1). The object header consists of the fixed-length "header core", which is mandatory, and the variable-length "header extension", which is optional and carries all information to identify and describe an object as well as some additional information for special user applications. Information carried in the mandatory "header core" is the size of the header and the body as well as the type and subtype information (e.g. type: image; subtype: JPEG). The optional "header extension" carries more detailed information on the object; some of the parameters carried there are described below.

The object body contains the data to be transported, for example a file.

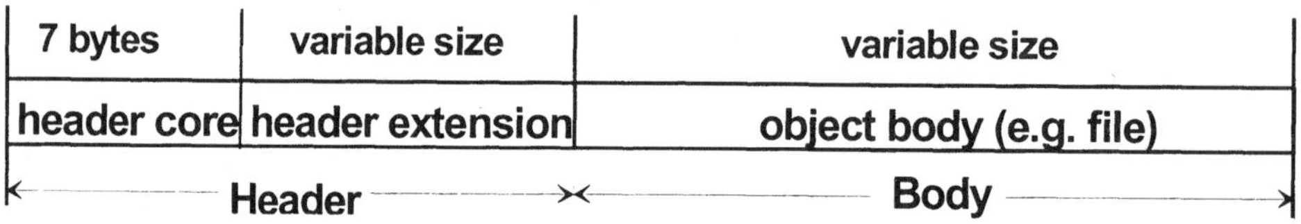

7 bytes	variable size	variable size
header core	header extension	object body (e.g. file)

|←—————— Header ——————→|←—————— Body ——————→|

Figure 4.1 General structure of MOT objects

For transportation the object is split into several segments. Each segment is mapped into one data group (for transportation details see below). This segmentation causes - together with the transmission of header information - some protocol overhead, but the segmentation is very important for error protection purposes: the smaller the segments, the lower is the error probability within a segment. Using repeated transmission at the data group level, the object can be reassembled from the error-free segments and the resulting error probability for the whole object is quite low. Currently data group sizes of about 1024 bytes are used as a good compromise between overhead and error probability.

During transportation the header information is separated from the body, either in an MOT header or in an MOT directory. This has the following benefits (among others):

- the header information can be repeated several times during the transmission of the body, for example if long objects are to be transmitted;

- the header information can be sent unscrambled when the body is scrambled for a user application using access control.

More information on the transport of MOT objects within the different DAB data channels is given in section 4.2.2.

4.2.1.3   MOT Header Parameters

Depending on the type and characteristics of the user application, the "header extension" may contain different parameters which the application needs for data management, presentation, etc. The parameters may appear in any sequence. This section describes some of the most important "header extension" parameters for MOT objects; a complete list can be found in (EN 301234).

ContentName. This parameter contains a unique name or identifier of the object and is used to refer to this object (together with the VersionNumber) unambiguously. Only one object with a given ContentName may exist within one user application.

The parameter identifies the one and only starting object within a user application of type "Broadcast Web Site" (see section 4.3.3). If the MOT directory feature is used, this parameter is only needed for compatibility with old data terminals.

TriggerTime. This parameter specifies the time, coded in short or long UTC format, when the object should be presented in a user application of type "Slide Show" (see section 4.3). If all information is set to zero, this means "Now" and the object should be presented immediately after complete reception.

4.2.1.4   The MOT Directory

For ease of storage and object management for the user application of type "Broadcast Web Site" (see section 4.3.3), the MOT directory has been defined. The directory is used to provide a complete description of the content of the "Broadcast Web Site" data set. Additional information on where to find the data for each identified object is also provided. In order to manage updates to the data set, for example the replacement or deletion of individual objects, version control mechanisms to both the objects and the directory itself are defined.

Within an MOT stream that contains a directory, there will be one directory that describes all objects within the user application data set. At most one directory is allowed within one MOT stream.

The MOT directory consists of the following parts:

Directory header. This header describes the size of the directory (length in bytes), the number of objects contained in the user application, the maximum time for one broadcast cycle of all objects contained within the user application, the default segment size for the MOT segmentation process (for data objects) and the length of the additional (optional) directory extension.

Directory extension. The directory extension carries a list of parameters which are used to describe all objects contained in the "Broadcast Web Site" data set. The structure and definition of these parameters is the same as for MOT objects.

Directory entries. The directory itself consists of a number of N (defined in the directory header) entries. Each entry starts with the "TransportId" which uniquely defines each object to which the following header refers. The attached MOT header consists of the header core and the header extension of the object and has exactly the same coding structure as for the header for object transport (data group type 3). It is not necessary that parameters of the header extension which are optional have to be carried both in the MOT directory as well as in the header extension of the object. In particular, to reduce the size of the MOT directory, only those parameter types relevant for object transfer should be contained within the MOT directory. If any parameters are present in both the MOT directory and the object header, the entries in both places have to be identical to ensure consistency of the parameters.

The MOT directory is carried within data groups of type 6.

Whilst older MOT data decoders did not support the MOT directory and identified the structure of the "Broadcast Web Site" data set by collecting the information needed from the object headers, it is highly recommended that this feature is supported in all future developments to ease the management of this kind of data service.

4.2.2   MOT Object Transport

4.2.2.1   Segmentation of Objects
As described above, the objects to be transported are split into segments to allow flexible handling of large quantities of data (see Figure 4.2). The header segments are structured

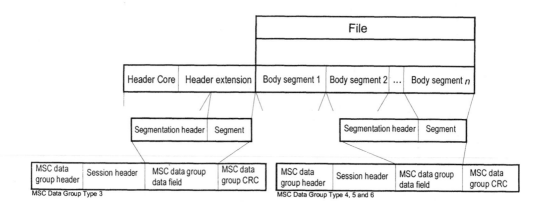

Figure 4.2 Segmentation of an MOT header and an MOT body

into data groups of type 3, whereas each individual segment will be mapped together with the associated "Segmentation Header" into an MSC data group data field. The body segments are treated the same way, but they are carried within data groups of type 4 or 5. MOT directories are transported using data group type 6. These data groups are transported in either one or more packets (if packet mode transport is used) or X-PAD sub-fields as described in the following sections. This transport will add another level of segmentation.

4.2.2.2   MOT transport in PAD

The MSC data groups are mapped 1 to 1 into an X-PAD data group. This data group will be split into X-PAD sub-fields for transportation. MSC data groups containing MOT data (data group types 3, 4, 5 and 6) are transported in X-PAD application types 12 and 13; in the case of CA messages (MSC data group type 1) the application types 14 and 15 are used. The basic principle of this mapping is shown in Figure 4.3.

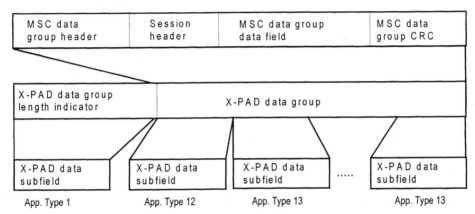

Figure 4.3 MOT transport in PAD

4.2.2.3   MOT Transport in Packet Mode

The MSC data groups are packetised as described in section 2.3.3.3. To distinguish between CA messages (see section 2.4) and data packets, the "Command flag" on the network level (i.e. in the packet headers) is used. This flag is set to "1" for MSC data group type 1, and to "0" for all other data group types (3, 4, 5 and 6) related to MOT. The basic mapping principle is shown in Figure 4.4.

4.3   Standardised MOT User Applications

Several basic end user applications based on the MOT Protocol have been defined and standardised. This allows data services to be offered which can be displayed by data terminals from different manufacturers.

M S C   D a t a   G ro u p   T y p e   3 ,4 ,5   a n d   6

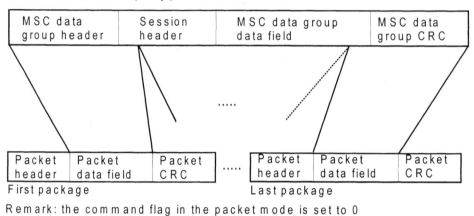

| MSC data
group header | Session
header | MSC data group
data field | MSC data
group CRC |

.....

| Packet
header | Packet
data field | Packet
CRC | ..... | Packet
header | Packet
data field | Packet
CRC |

First package   Last package

R em ark: the com m and flag in the packet m ode is set to 0

Figure 4.4 MOT transport in packet mode

4.3.1   Asynchronous MOT Slide Show

An MOT Slide Show (TS 101499) consists of a sequence of objects which should be presented to the user automatically without interactivity. This presentation can be done either cyclically or in a single sequence. The Slide Show usually consists of a sequence of (JPEG or PNG) images.

All objects are broadcast in the same sequence as they should be presented to the user. To identify these objects as part of the Slide Show, their header extension must contain the parameter "TriggerTime"; the value is set to "Now" (all zeros) for asynchronous mode. If the corresponding service is selected at the data terminal, every object will be presented immediately after the complete reception and a possible decoding delay.

If the object transmission time is much lower than the time interval between the presentation of different objects, a repeated transmission of an object (object repetition) is possible. This reduces the delay until the first presentation of an object after switching on a data terminal or selecting the service carrying the Slide Show. For such a repeated object transmission all parameters should be the same during the repetitions. A terminal which has already received and presented this object should not react again on this object.

If a new object is to be presented, either the ContentName or the VersionNumber must be changed at the broadcasting side. This forces a change of the TransportId which triggers the terminal to receive the new object.

With respect to the implementation of the Slide Show service two major aspects have to be taken into account:

- for current data terminals the decoding delay for JPEG images is in the range of 2...20 seconds depending on the size and complexity of the image;

- the minimum distance between two objects in an MOT Slide Show targeted at mobile users should be at least 30...60 seconds.

4.3.2   Timed MOT Slide Show

This user application is quite similar to the asynchronous Slide Show, but now each object will be presented at a pre-defined time which is specified by the parameter "TriggerTime". Care must be taken at the broadcasting side that the time reference used for service generation is the same as for the time transmitted within the FIC to guarantee a synchronous presentation. In addition, it should always be guaranteed that the object to be presented next is completely received at the terminal before "TriggerTime" minus the decoding delay for a synchronous presentation.

For a timed MOT Slide Show the objects to be presented should be broadcast some time in advance of the time when they should be presented to ensure that they are completely received before the trigger time occurs. The received objects are stored at the data terminal, but not resorted. In order to keep the storage requirements for the data terminals low, the service provider should keep in mind that simple receivers can only store up to two objects in addition to the object actually shown and design the service accordingly.

The interval between the presentation of two objects should be large enough to receive and decode all consecutive objects and not to gain too much of the car driver's attention, if the service is targeted at mobile reception. As for the asynchronous Slide Show, it is advisable to have a minimum time of 30...60 seconds between the presentation of two objects. Before establishing such a timed MOT Slide Show it should also be checked which time resolution is used in FIC signalling (the short format has a resolution of only 1 minute, see section 2.5.2.2) and whether the data terminals support fine time resolution to the second (some data terminals are able to generate the seconds resolution on their own).

The feasibility and application of such a timed MOT Slide Show application have been demonstrated at exhibitions (Hallier, 1994b) and in some pilot projects, but currently no such service appears to be permanently available.

4.3.3   Broadcast Web Site

The user application "Broadcast Web Site" (TS 101498) offers the user an Internet-like data service with local interactivity. Such a service consists of up to several hundred linked objects (HTML and image files, e.g. JPEG) which are organised typically in a tree-like structure below a single start page. The objects of such a service are broadcast cyclically in a "data carousel". To minimise the start-up time for a user who turns on his or her terminal or selects the "Broadcast Web Site" application, the data objects of the application should be prioritised and broadcast accordingly. It is advisable to broadcast the start page very frequently, say every 15...30 seconds. The next level of objects - normally the objects which are directly referenced by the start page - should also be broadcast rather frequently, the next levels less frequently and so on. Also the priority of larger JPEG pictures could be handled independently of the HTML pages; this can further improve the response time. This priority management has to be done at the data multiplexer and is for example implemented in the data server of one supplier (Fraunhofer Institute). Using such a priority management, the user may start to navigate through the upper level of information already received while the lower levels are received in the background.

It is possible to actualise single objects or parts of the application simply by transmitting a new object carrying the same ContentName with a new TransportId, for example for updating traffic information or news. Some data terminal implementations (e.g. the Grundig terminal) react to such an update with an immediate redrawing of that object if it is currently presented to the user. This can be done either during the normal transmission of an object or later on by a "header update".

Using the MOT directory, the object management is quite easy: having received the directory, the terminal can decide for example how to organise the storage, which objects should be received or discarded, etc. The updating and deletion of objects can now be explicitly controlled by the service provider broadcasting an update of the MOT directory. The current drawback of the MOT directory is that there is only a very limited number of data terminals supporting this feature, but the next generation of data terminals is supposed to support MOT the directory also.

4.3.4   Interactive Services

DAB data-casting can be made interactive by adding a second bidirectional channel, such as provided by a cellular telephone system (e.g. GSM). This was first investigated and demonstrated in the context of the "MEMO" (Multimedia Environment for Mobiles) project within the ACTS programme of the EC (Lauterbach, 1996; 1997a; 1997b), (Klingenberg, 1998).

Since mobile telephone systems today only allow for restricted bit rates, this combined system provides an asymmetric link with a high download capacity on DAB which is well suited for the following applications:

* Wireless Intranet access, for example access to databases for travelling salespeople and emergency services
* Wireless Internet access for professionals (journalists, construction engineers), etc.
* Entertainment and travel information in public means of transport.

Figure 4.5 shows the MEMO system layout using DAB and GSM. The data flow is as follows: the user terminal which needs access to a particular server of the Internet sends its request to the gateway server via GSM, which forwards it to the service provider.

Depending on the amount of data to be delivered to the mobile terminal, the data will be sent via GSM or DAB. The gateway server will indicate this to the terminal and also convert the data to the respective protocols used on the different networks (e.g. IP, MOT).

In the context of MEMO and Eureka 147 a number of standards have been developed which support this kind of interactive use of DAB with other bidirectional systems. Among these are (ES 201735), (ES 101736), (ES 201737).

After the termination of the MEMO project the "M4M Forum" (Multimedia for Mobiles Forum) was founded to further promote the integration of broadcasting and tele-communications systems and to provide early UMTS-like services (www.m4mforum).

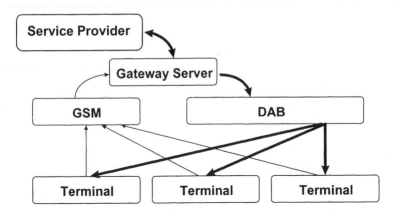

Figure 4.5 MEMO system concept

4.4 "Mobil-Info" as an Example of a Proprietary MOT User Application

The data-casting facility of DAB and MOT can be used to build a bunch of mobile multimedia presentation applications and kiosk systems. One example of such a service is "Mobil-Info". This service was developed by a German network and service provider (Deutsche Telekom AG Broadcasting and Trend Network AG), see (www.mobil-info).

The main idea of the service is to combine DAB data-casting with a hard-disk-based video presenter. Beneath video playback the PC-based presenter system merges different media like text, still picture and script-based sprite animation into an appealing programme. The presenter system can be installed for example in trains and buses in public transportation services. Actual information that people want to receive immediately is transmitted via a small and cheap DAB data channel. The presentation is enriched with graphics and video components that are stored on the hard disk of the system. High-resolution graphics and video that would need high transmission capacity are downloaded to the system once a day over a wireless LAN in the railway-yard, for instance, or they are transmitted during the night when less DAB capacity is used for radio programmes.

The video clips and the script-based graphic animations may be trailers for public services, cultural events or advertisements. While video clips would require an un-acceptably long transmission time, the size of the script animation can be kept very small. A script can compose different multimedia objects on the fly. Some of these, such as background pictures, may already be stored on the hard disk while others are transmitted via DAB. The composing of animation scripts on the fly allows actual information with high resolution to be combined with good-looking graphical content. The video and script-based graphic animation are presented in full-screen mode and are controlled by a play list. The entries of the play list can have different priorities, which is important for trading with advertising time slots. The default play list, downloaded via the wireless LAN, can be changed during mobile operation via DAB by special video presenter control commands. Actual information like traffic messages or football results can be easily inserted in the programme.

In addition to the play list an information channel is available. It is represented as a sliding news ticker. The news ticker runs asynchronously to the play list and is made up completely of news transmitted over DAB. The overall structure of the service is shown in Figure 4.6.

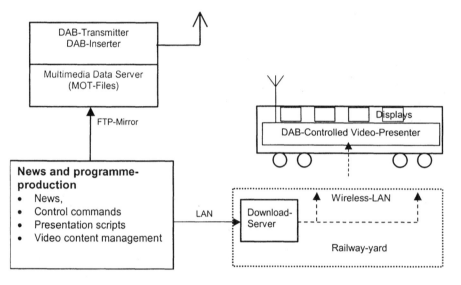

Figure 4.6 Overall structure of the "Mobil-Info" service

The trains in the service are equipped with the DAB-controlled video presenter. Outside the railway-yard the video presenter receives information via DAB. Inside the railway-yard the video presenter communicates with a local server via the wireless LAN. The larger part of the information to be stored on the trains is downloaded to the railway-yard server over a private local network connection from the news and programme production service centre. The service is based on an off-line and an on-line production. In the off-line production process the programme is created, time slots for advertising can be booked; video clips are converted to an appropriate format; graphical animations are created; and play lists are generated. The on-line production organises actual incoming information. This actual information may be submitted from the public service provider or from other sources. The news ticker is fed with actual news, and can be updated within a few minutes. The full screen can also be used for actual information inserted by DAB. The play list contains some entries with lower priority. These clips can be shortened to allow the introduction of actual information in the form of graphical animations. For example, a weather report or the results of football matches can be presented with an animated sprite graphic. The on-line production has a connection to the DAB inserter. This is realised as an FTP mirror process.

The DAB-controlled video presenter is realised with standard PC components. It uses a DAB receiver card and a powerful video system. It also has an interface to the information bus system of the train. This is used to get the names of the stations and to display them on the screen. In this context the DAB interface may be of some interest to readers. It is built on top of the MOT Protocol. The MOT objects RAW, JPEG, GIF, MPEG-1 and MPEG-2 are supported. The control mechanism of the video presenter is implemented as a proprietary protocol on top of the RAW data object. A set of control commands are defined which are used to feed the sliding news ticker, to control the play list and to transmit systems commands. Each mobile system can be addressed by an individual system identification command. An addressing scheme similar to the Internet scheme is used. Groups of systems can be addressed globally, individually or by multicast. The addressing scheme allows the content and information shown to be tailored to the local requirements. For example, one advertisement or one piece of actual information can be shown only on a specified train line.

The set of news ticker commands allows news to be added and removed from the news ticker, to turn it on or off and to influence the speed and the colour. The play list commands are used to add new entries to the play list. The animation script, the GIF and JPEG files or the video used for this new entry can be transmitted by DAB, too. The control commands support a mechanism that guarantees that all elements required for the new entry are received before starting the presentation. Though the downloading of video is possible, the concept of "Mobil-Info" would not recommend this. The service is designed for DAB data-casting applications that use only modest data transmission rates. Videos would therefore require quite a long time to download. For example, in a DAB channel with a capacity of 12 kbit/s a video with a length of 15 s and a bit rate of 5 Mbit/s (75 Mbit) would take approximately 2 hours to download. On the other hand, downloading a picture of 200 kbytes (2.6 Mbit) would take only 3 minutes.

The "Mobil-Info" service is currently running as a pilot project in Potsdam, Germany. Four trains are equipped with a video presenter, each supporting four double-page displays. Advertisements are merged with local information and actual news. The acceptance of the service among the population is high and the forecasts promise good commercial success.

4.5   Textual PAD Services

4.5.1   Dynamic Label

The DAB Dynamic Label user application is comparable to the "RDS Radiotext" known from analogue FM broadcast. The basic functionality is to output short textual messages, the so-called *labels*, on a display. These labels are segmented during transport in up to eight segments each consisting of up to 16 characters. The segments are transported in X-PAD data groups. These data groups carrying Dynamic Label segments are signalled with the X-PAD application types "2" (start of data group) and "3" (continuation of data group). The basic structure of an X-PAD data group carrying a Dynamic Label segment is shown in Figure 4.7.

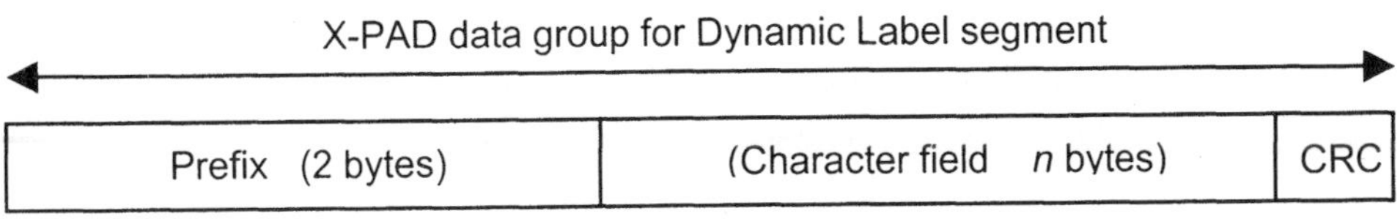

Figure 4.7 X-PAD data group carrying Dynamic Label

Each X-PAD data group carrying a Dynamic Label segment consists of a 16 bit prefix with control information, followed by a character field of up to 16 characters as part of the complete label string. At its end the data group carries 2 bytes (16 bits) of CRC information for error detection.

The segments of a single Dynamic Label string can be transmitted in any order, where repetitions of segments can be used to increase the reception probability in case of high bit error rates in the transmission channel, but it is not allowed to interleave segments of different label strings. The segments can be identified by the segment number carried in the prefix field.

For adaptation to the needs of foreign languages the character set to be used for displaying the label is specified in the prefix field also. Currently five different character sets are identified, whereas the default character set comprises the complete EBU Latin-based set; other sets contain a common Latin-based core and Greek, Cyrillic, Arabic or other extensions. Current implementations usually support the complete EBU-Latin based set only.

In order to allow some minimum formatting capabilities for the labels and to optimise the label presentation for smaller displays, some special control characters are defined. Using these control characters, preferred word and line breaks may be defined as well as a headline feature. In addition, a control code to remove any label from the display is provided. The additional control codes count as normal characters; therefore the maximum number of 128 displayable characters within one Dynamic Label string will be reduced by the number of control characters used within all segments of one label string. For more information regarding the coding and formatting of Dynamic Label strings see section 5.4.4.

Typical applications for Dynamic Label are news flashes including the weather forecast, title and interpreter of music currently played, details of actual events or even commercials. Common to all these services is the close relation to the audio programme; therefore this kind of service is carried within the PAD channel.

4.5.2   Other Textual PAD Services

For the DAB system, ITTS (Interactive Text Transmission System) is defined as an alternative mechanism to transport textual information to data terminals. For transporting the ITTS data streams either X-PAD can be used, or for non-directly audio-related services stream and packet mode sub-channels may also be used.

The ITTS standard (ITTS, 1994) offers many different possibilities for text formatting and controlling the presentation of the text, including basic block graphics similar to Videotext used in TV. In addition, several different text channels can be multiplexed and selected with an appropriate user interface. The layered structure of this service allows at the receiver side a flexible adaptation to different types of displays – from monochrome single-line up to colour graphics terminals.

ITTS is derived from the "Digital Compact Cassette" standard where it is used for additional textual information. In practice, ITTS no longer has any relevance for DAB since the decoding is rather complicated and basically all features offered by ITTS could be realised by using HTML also. Therefore, for textual PAD services requiring more formatting capabilities and longer texts than supported by Dynamic Label, MOT services carrying one or more HTML pages ("Broadcast Web Site") are used.

4.6   Traffic Information Services and Navigation Aids

In this section two approaches to traffic and travel information (TTI) services and support for differential GPS based on DAB are described. For a broader view on travel information broadcasting see (Kopitz, 1999).

4.6.1   Traffic Message Channel (TMC)

"Traffic Message Channel" (TMC) is a basic data service consisting of digitally coded traffic messages, and·was originally defined for RDS (Radio Data System, EN 50067) services in FM broadcast. All messages are coded according the "Alert C" Protocol and carry the following information:

- characteristics of the event: detailed information describing the traffic problem or weather situation;
- location: area where the problem occurred;
- severity of the event;
- duration: estimated duration of the event;
- alternative route.

All this information is given in numeric code values which refer to the various tables (event, location, direction, duration, etc.) of the Alert C Protocol; these code values are assembled in "groups". Each group consists of up to 37 bits of information.

Additional information can be given with special messages, e.g. a more precise definition of the event's location or a description of an exceptional event not covered by the standard message table.

The handling of the messages (validation, time-out processing, priority management, etc.) as well as the presentation to the user could be done by an external decoder. The presentation can be done either by an alphanumeric display or by a speech synthesiser which generates spoken messages.

The TMC messages are carried in the FIC in FIGs of the type 5/1. Several messages may be assembled within one FIG 5/1 data field.

A traffic information service based on TMC carried on DAB was first demonstrated in 1995 when an external TMC decoder box which was originally designed for FM-RDS TMC was connected to the serial port of a DAB data decoder. Since the infrastructure for collecting the traffic information data was rather poor at that time, the service was for demonstration only. In addition, the real-time insertion of the TMC messages into the FIC was rather complicated. During the last few years this infrastructure has improved significantly and therefore TMC might be used for setting up a useful traffic information service. Also the capabilities of the multiplexers and the data inserters have been improved with the introduction of the "Service Transport Interface" (STI) which allows easy insertion of TMC messages assembled in FIG 5/1.

4.6.2   TPEG

TPEG stands for the Transport Protocol Experts Group of the EBU which developed an
end-user-oriented application for the delivery of road traffic messages (Marks, 2000). The
specification defines the core protocol, the network and the service layer. Service providers
may operate services which may use one or more delivery technologies (e.g. Internet, DAB,
etc.) from one message generation process. The application layer is scalable and allows a
range of receiver types to be used simultaneously – from simple receivers with only a small
alphanumeric display up to integrated navigation systems. One of the key characteristics of
TPEG is the fact that there is no need for the end user to have a location database before
using the service. The location database is derived from digital maps at the service provider
side and the location information will be transported to the end user embedded in the TPEG
application.

TPEG is organised as a self-consistent stream of data which contains all messages as
well as provisions for synchronisation and error detection. With these characteristics, a
TPEG stream should be broadcast over DAB using the "Transparent Data Channel" (TDC)
specification (TR 101759), see section 4.8. This results in carrying the TPEG data stream in
a completely transparent way over a virtually stream-oriented channel – at the receiver side
the bytes come out in the order they go in at the transmitter side.

First implementations of TPEG will use an external application decoder connected to a
DAB data receiver and will be available as prototypes at the end of the year 2000.

4.6.3   Differential GPS

For navigation purposes the satellite-based "Global Positioning System" (GPS) is widely
used. A miniaturised satellite receiver receives information from five or more satellites and
calculates from the data sent by the satellites and the receiving delay its current position.

The remaining positioning uncertainty with "plain" GPS is in the range of
$\pm 25...100$ m. In order to reduce this uncertainty, "Differential GPS" (DGPS), can be used.
The basic principle is to use a stationary reference receiver with a well-known fixed
location. This reference receiver compares its known location with the location calculated
from the satellite data. The difference between these two locations is transmitted in the
standardised "RTCM 2.1" format via a broadcast channel to mobile GPS receivers. These
receivers may then correct the position calculated from the satellite data by the additional
information and so reduce the positioning error by a factor of $5...10$.

Based on the principle described above, a DGPS data service has been implemented and
is still operated in Bavaria, Sachsen-Anhalt and also in northern Italy. A GPS reference
receiver is connected to a serial input of the DAB data multiplexer which collects the
RTCM data. Each RTCM data record is assembled into one MOT data group and
transported with the ContentType/ContentSubtype "proprietary" since no entry was defined
in the related MOT tables at service definition. The DAB data receiver decodes these
records and sends the RTCM data as a serial data stream to an RS232 output which is
connected to the mobile GPS receiver.

For actual implementations use is recommended of the "Transparent Data Channel"
(TDC, see section 4.8) feature of DAB which offers a "virtual serial data connection"
between the transmitting and the receiving side. In addition, in the meantime DGPS is

included in the MOT tables and also defined as a user application in the service signalling mechanism base of FIG 0/13 (UserApplicationId 4).

4.7   Moving Picture Services

4.7.1   General

The DAB system has proven its superb suitability, especially for mobile reception. Shortly after demonstrating the first multimedia data services as described above, an idea arose to add to DAB the ability to transmit moving pictures. This was not intended to concur with DVB, but as a supplement to the pure audio and basic data transmission to be used for special applications and niche markets.

One of the basic problems to solve is the very high data rate required for video transmission: without compression these data rates of the video signal are in the range of up to several hundreds of Mbit/s, depending on the target screen size, resolution and colour depth. To reduce these data rates to a region more applicable for transmission and storage, several compression algorithms have been developed and standardised. One of these standards is MPEG-1 (IS 11172) which requires a data rate of about 1.5 Mbit/s for transmitting video signals comparable to PAL resolution and quality. Another coding algorithm, targeted more at very low bit-rate applications like video conferencing on ISDN lines, video telephony and the like, is the ITU-T standard (H.263).

Based on these different coding strategies, there were mainly two different approaches for video broadcasting using the DAB standard: one proposal, called "Digital Multimedia Broadcast" (DMB) is mainly based on MPEG-1 coding and focuses on TV-like applications, whilst the "Motion PAD" (M-PAD) was mainly designed as a supplement to the audio programme, which offers a slow-motion, low-resolution H.263-coded video in parallel to the normal DAB audio programmes. These two systems are characterised in the following sections.

4.7.2   Digital Multimedia Broadcast (DMB)

The DMB system can be characterised as digital TV broadcast based on the DAB transport mechanisms. The video and audio signals will be source coded according to the MPEG-1 standard (IS 11172) using target bit rates in the range from 1.2 up to 1.5 Mbit/s, adapted to the "Ensemble Transport Interface" (ETI) (EN 300799); see also chapter 6. They are fed directly to the DAB ensemble multiplexer/modulator. Since a DAB ensemble may carry about 1.5 Mbit/s of useful data (depending on the code rate), one MPEG-1 bit-stream will roughly fill one complete ensemble and leave only a small capacity for use by other applications.

At the receiving side, a standard DAB receiver is used for channel decoding up to the serial data interface "Receiver Data Interface" (RDI) (EN 50255), see also section 8.5. Owing to the high bit rate used, a "full-rate" RDI interface will be necessary. The RDI data stream is then fed to an MPEG-1 decoder which outputs the audio and video signals for presentation. A schematic of the general system structure is shown in Figure 4.8.

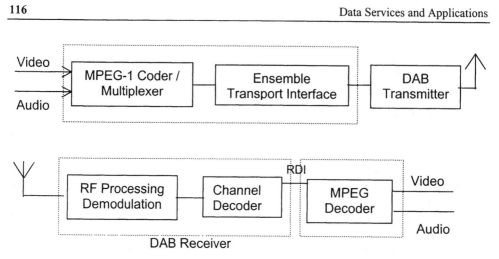

Figure 4.8 Schematic of a DMB system

It was shown in several demonstrations and field trials that this kind of TV transmission to mobile receivers dramatically improves the reception quality in comparison to conventional analogue TV broadcasts (Hallier, 1995).

Currently there are implementations of DMB broadcasting TV programmes to high-speed trains in Germany and Sweden; another implementation offers a travel information service for public transport in the Hanover area. According to (Bosch, 1999), 144 trams are equipped with double screens on which travel schedules, connections, delays and updates are displayed, together with video spots and commercials. The Hanover area is covered by three transmitters with a power of 200 W each and 13 repeaters are used to cover tunnels and underground sections.

4.7.3   Motion PAD (M-PAD)

The main intention during the development of the M-PAD service was not to define a new video transmission system, but to enhance existing audio services by accompanying the audio with visual information. Therefore the following goals have been defined:

- M-PAD video must be a compatible extension to existing audio services;
- the video signal should be transported in the PAD section of the DAB audio since the video is strongly related to the audio;
- due to the PAD transport the bit rate used for the video is limited to 64 kbit/s;
- the video should be synchronous to the audio.

To match these requirements, the ITU-T standard (H.263), mainly designed for video telephony and video conferencing systems, was chosen in its enhanced "Version 2" as the coding algorithm for M-PAD. While this coding algorithm basically only defines the bit-stream format and the decoding method, for the encoder some parameters may be fine tuned according to the application. So the actual bit rate or image quality can be controlled as well as the frame rate and some other system parameters. The following basic parameter values have been chosen:

- image resolution: 176×144 pixels,
- bit rate for the video: 64 kbit/s.

For the transport the encoded video stream is segmented into self-consistent blocks, which can be decoded without any cross-relations to the previous or the next block. The block length has been adjusted to a playback length of about 3 seconds. This length is a compromise between coding efficiency and delay both for resynchronisation in case of reception errors as well as for the initial synchronisation of newly tuned receivers.

These blocks are transported within the X-PAD of the audio as individual MOT objects according to the DAB and the MOT standard.

Inherent to all source coding algorithms, especially the high compression ones, is the *coding delay*. Since this delay cannot be avoided, the audio must be delayed at the transmitting side at least by the delay caused by the overall video transmission chain (video coding, MOT segmentation and transport, MOT decoding, video decoding). This delay of the audio signal is controlled by the M-PAD encoder, which is located between the audio encoder and the DAB multiplexer, as well as the synchronisation between the audio and the video. This synchronisation is done as follows:

- The coarse control of the playback time of an individual H.263 video object is done by the MOT parameter "TriggerTime" which controls the presentation time of an object exactly to the second (in relation to the DAB system time transmitted within the FIC).
- The fine synchronisation is done by extra sync tags which are carried within the F-PAD as 6-bit "In House Information". With these sync tags the synchronisation between the audio and the video can be done exactly to the millisecond.

The M-PAD system was been first demonstrated in the German DAB project Sachsen-Anhalt (Halle) and has been available there as a regular service since 1999. For decoding and presenting the video part of the service, special data terminal implementations are required. Currently there exists one special device, the mobile DAB multimedia terminal "DAB M³", which is capable of decoding audio services including Dynamic Label, MOT Slide Show as well as Broadcast Web Site, and in addition has an M-PAD decoder. Another device supporting M-PAD is a DAB PCI card with special decoding software.

4.8   Transparent Data Channel

The "Transparent Data Channel" (TDC) (TR 101759) provides DAB with the ability to transport serial data streams using the following DAB data channels:

- TDC in a packet mode service component;
- TDC in a stream mode service component;
- TDC in an audio service component carried as X-PAD.

These serial data streams are transported from the transmitter to the receiver in the order in which they are fed to the transmitter. There is no implied beginning or end to the data and therefore any "framing" and/or synchronisation of data within the stream must be handled by the application protocol which is carried by the stream. The TDC Protocol transmitting serial data streams can be regarded as a complement to MOT which is mainly intended for object (file) transfer.

Because the reliable delivery of data cannot be assumed for a broadcast channel in the mobile environment, and since there is no return channel for granting reception or requesting a retransmission of the data, applications using the TDC for data transport must ensure that they are tolerant to errors in the data. This applies especially to loss of data and corrupted data.

Typical applications using the TDC protocol are DGPS or TPEG.

5

Provision of Services

THOMAS SCHIERBAUM, HERMAN VAN VELTHOVEN and
WOLFGANG HOEG

5.1 The DAB Service Landscape

The introduction of DAB took place during a period of great change in the radio world, in general. Following the digitalisation of studios and consumer products, now the means of transmission to living rooms and cars are going to be digital. Furthermore, in addition to speech and music, till now the only forms of presentation to audiences, new kinds of multimedia applications will be introduced with the Internet, digital broadcasting and new generations of mobile communication systems.

The migration from the existing to necessary new infrastructures for the provision of multimedia services is a major challenge for content and service providers. Depending on the intended amounts of investments, the appearance of service providers in DAB will be very variable. Most of the radio stations are starting with a simulcast of existing analogue audio services. Some stations are also providing data services for a better general service for the audience. However, few providers are launching new audio services. In all these cases, the accessibility of listeners plays a decisive role. As a result of the normally slow appearance of new technologies, broadcasters have to exploit synergetic aspects of other media during the generation of a service. One time generating – several times distributing.

The intention of this chapter is to describe the service generation, the adaptation of the DAB system to interfaces of other media and the gradual integration of the necessary infrastructures in broadcasting houses. Most of the examples described are taken from the experiences of German broadcasters.

Digital Audio Broadcasting: Principles and Applications, edited by W. Hoeg and T. Lauterbach
©2001 John Wiley & Sons, Ltd.

5.1.1   Structure of DAB Service Organisation

In the past, only a simple structure of responsibility was needed in a broadcast system: the programme provider (editor, supervisor or similar) was responsible for the production of the final content and form of a radio programme and for the studio output into the broadcast chain for distribution and transmission to the customer, without any further changes in content or quality. This was true for AM and FM/VHF radio, and partly also for the first digital radio services (DSR, ADR, etc.).

The very complex structure of the contents of DAB (i.e. audio programmes with different formats and quality levels, programme associated data, independent data services, programme service information) and its grades of freedom to change dynamically several parameters of the multiplex requires a more diverse responsibility for managing the final content, form and quality of a DAB service by

- the programme provider
- the service provider
- the ensemble provider.

Figure 5.1 shows a simplified structure of the processes of service provision for a DAB service ensemble.

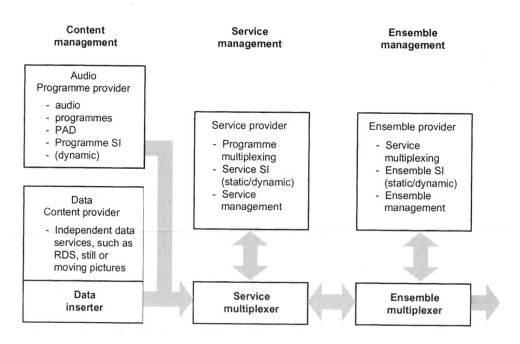

Figure 5.1   Structure of the management process of DAB services

Programme provider

For audio programmes, the programme provider may have the same tasks concerning content as in a conventional audio broadcast service, but, in addition will also provide some other programme-related information (Programme-associated Data PAD) such as DRC (Dynamic Range Control) data, M/S (Music/Speech) flag, programme SI (Service Information) such as dynamic label, language assignment (see section 2.5.4.1), programme type assignment (see section 2.5.4.2), announcement type (see section 2.5.4.5) and so on. All these data are of dynamic type, which means that they will change in a close time relationship to the current programme content.

In a similar manner the programme (or content) provider for independent data services will manage the contribution to a certain DAB service. This can be located apart from the broadcast studio. The audio coding process itself can be allocated either in the broadcasting operating system or in the service multiplexer, which is the central component of the service management process.

Service provider
The programme output is passed to a service provider, and a series of programmes are then assembled into a programme service which may also contain other components in addition to the audio programmes, for instance SI similar to RDS for FM/VHF broadcasting, or new, independent data services.

The SI can be of static type (such as programme service label, see section 2.5.3), programme type (see section 2.5.4.2), regional identification (see section 2.5.5.4), or of dynamic type (programme type preview (see section 2.5.4.4), programme number (see section 2.5.3.4). Also new data services (e.g. traffic information or still pictures for different purposes) which are completely independent of the main (audio) service can be assembled in a service multiplex. The service provider's task is to define the organisation of the service (i.e. the description of each service component and how they link together), see (BPN 007).

Another task of the service provider will be the management of the service multiplex, such as reconfiguration requests or changes in audio parameters, such as data rate, mono/stereo and so on. Each digital audio bit stream has an individual data rate and audio type (mono, stereo, Joint Stereo, etc.).

Ensemble provider
Finally, a number of programme services (about four to seven), including the other components of the Main Service Channel (MSC) and the Fast Information Channel (FIC), are multiplexed by the ensemble provider in one DAB service ensemble and finally fed to the DAB transmitter(s). The ensemble provider is also responsible for particular static and dynamic SI as shown in Figure 5.1. In general, this task will be managed by a separate organisation apart from that providing programme content or services.

More details of selected aspects of the provision of services and the necessary infrastructure of the production processes for DAB are given in the following sections. As these technologies are still not very common at the time of writing, the descriptions may be understood as an overview by means of examples or proposals for future implementation.

See also Chapters 2 (system concept), 3 (audio services and applications), and 4 (data services) for details of the service components mentioned. Chapter 6 (collection and distribution networks) details the functionality of the service and ensemble management processes and the interfaces between them.

5.1.2   DAB Main Services

This section explains the structure of DAB services in some detail, for a better understanding of the specific abstracts and coherence used later in the text.

Audio services, data services and the necessary configuration information are assembled as a so-called DAB multiplex. This multiplex is configurable and provides a data rate of approx. 1.2 Mbit/s. It is designed to transport an FIC and an MSC including the individual sub-channels which transport the audio and data services.

The DAB MSC provides mainly four service transport mechanisms:

- Service Information (SI),
- Sub-channel for audio services
- Programme-associated Data (PAD) carried within audio services
- Packet mode data carried in individual sub-channels for data services.

An example of a multiplex configuration with multiple audio services used in practice is shown in Table 5.1.

Table 5.1   DAB multiplex configuration for Ensemble Bavaria, Germany (Channel 12D)

No.	Service	Service ID (Hex)	Audio Mode	Bit rate (kbit/s)	SubCh ID (Hex)	SubCh Size (Cu)	Error Prot.
1	Bayern 2 Radio	D312	JTStereo	160	2	116	UEP3
2	Bayern 3	D313	JTStereo	160	3	116	UEP3
3	Bayern 4 Klassik	D314	JTStereo	192	4	140	UEP3
4	Bayern 5 Aktuell	D315	Mono	96	5	70	UEP3
5	Bayern Mobil	D316	JTStereo	160	1	116	UEP3
6	Rock Antenne	D319	JTStereo	192	7	140	UEP3
7	Radio Galaxy	D31B	JTStereo	160	6	116	UEP3
8	Mobile Data BDR	E0D01008	none	64	11	48	EEP4

Note: UEP = Unequal Error Protection level, see section 3.7.

Sub-channel for audio services
The DAB sub-channels allow the transmission of audio signals according to the audio standard ISO/IEC MPEG-1 Audio (IS 11172) or MPEG-2 Audio (IS 13838). The transport capacities are ordered from content providers at the network providers in the form of Capacity Units (CU). The number of CU defines the audio bit rate or packet mode bit rate at the chosen error protection level (i.e. 140 CU; protection level 3 = 192 kbit/s). For considerations of service variety and audio quality the transport of approx. six to seven audio services per multiplex are possible (see Table 5.1).

Programme-associated data (PAD)
Parts of the audio sub-channel capacities can be used for the transmission of ancillary data. The information is transported synchronously within the MPEG-1 audio bit stream. The PAD comprises a Fixed PAD (F-PAD) control channel with a data rate of 0.7 kbit/s and an optional Extended PAD (X-PAD) transport channel with capacities in principle up to 64 kbit/s. Most of the actually offered audio encoders have a restriction of 16 kbit/s.

The following information can for instance be transported in PAD:

- F-PAD:
  - Size of the extended data transport channel (X-PAD)
  - Control information for different listener situations DRC, M/S flag, see Chapter 3
  - Ordering information (European Article Number, EAN, International Standard Recording Code, ISRC)
  - In-house Data.
- X-PAD:
  - Dynamic label (DAB Radiotext)
  - Multimedia Object Transfer (MOT) protocol, for details see Chapter 4
  - Transparent Data Channel, TDC, (in standardisation)
  - Transport Experts Protocol Group, TPEG, (in standardisation).

Owing to the necessary time synchronisation of audio signals and the associated data, the PAD insertion should be operated under the responsibility of the content providers.

As a result of the synchronous transport of audio and PAD in a common bit stream arises following problem. During the audio encoding process the available bit rate results from the selected audio data, determining the PAD data rate. In cases of low audio bit rates, for instance 96 kbit/s, and a maximum PAD data rate of 16 kbit/s, a reduction in audio quality may occur. The remedy is a sensible configuration of the capacities for audio and PAD data.

Packet mode data
Besides the audio services, additional sub-channels can be configured in the MSC for the transport of packet mode data services. Under the primary aim of covering the existing analogue radio market in DAB during introduction of the system, capacities of approx. 64 kbit/s for packet mode data services were realistic. With additional DAB frequencies in the future, higher data rates can be feasible for multimedia or telematic services. The packet mode data are to be inserted either at the locations of the service providers (into the service multiplexer) or at the network providers (into the ensemble multiplexer).

5.1.3.   Data Services

5.1.3.1   Dynamic Label
The Dynamic Label format transports transparent text information and control characters with a length up to 128 characters within the PAD channel. The service can be easily presented with alphanumeric text displays and thereby readily realised with cheap DAB receivers. The receiver supports presentation of text according to the implemented display type, that is 32 characters per two lines or 64 characters per four lines. The first receivers with possibilities for incremental or scrolling functions are available on the market. The broadcasters are responsible for a sensible service duration of single Dynamic Labels for text presentations in moving cars, see also section 5.4.4.

5.1.3.2   Multimedia Object Transport (MOT) Protocol
The MOT protocol allows the standardised transport of audio visual information, such as still pictures and HTML content. The use of MOT is similarly possible in the PAD and

packet mode. Two applications of the MOT protocol are frequently used in DAB projects: the service *Broadcast Web Site, BWS (TS 101498)* and the service *Slide Show, SLS (TS 101499)*.

5.1.3.3   Broadcast Web Site (BWS)

The BWS is a local interactive service, where the user selects the information already received with a browser. This form of a "radio web" bases on the mark-up language HTML. Besides the application, profile types are also fixed in the standard. The profile types rule the technical requirements of the presentation platform. Two profiles are associated with the BWS service. One is for services at integrated DAB data receivers, that is for car PCs or navigation systems, with a display resolution of ¼ VGA (320 x 240 pixels), HTML ver. 3.2 and a storage capacity at least of 256 kbytes. The second profile allows a non-restricted service presentation, that is on PC platforms. Supporting the first profile has resulted in a larger accessibility for users, because the ¼ VGA profile can be received at PC platforms as well as at integrated receivers.

5.1.3.4   Slide Show (SLS)

This second application describes on the basis of JPEG or PNG (Portable Network Graphics) files sequences of still pictures. The order and presentation time of this service are generated by the provider. This service provides no local interactivity to the user. The transmission time depends primarily on the file sizes of the pictures and the PAD data rate (see Table 5.2).

Table 5.2   Transmission time of MOT services (examples)

Content	Number of Files	File Size	Transmission Time (Data rate = 16 kbit/s)	
			PAD	Packet Mode
CD-Cover (JPEG) Resolution 320 x 240	1	14 kbytes	7 s	7 s
CD-Cover (JPEG) Resolution 640 x 480	1	42 kbytes	22 s	22 s
HTML file Text only	1	1 kbyte	0.5 s	0.5 s
HTML files	37	129 kbytes	1.10 min	1.05 min

5.2   Use of Existing Infrastructures

The radio world is going to be digital. This statement is valid for radio waves as well as for audio signal recording and transmission and the work flows in the broadcasting houses. The introduction of computer-aided radio has lead to more efficient and simplified work flows. Computer modules connected over digital networks satisfy the requirements at a higher rate during information processing, easy format checks and faster programme production. The usage of computer-aided radio technologies provides a ready source for new services.

5.2.1   Broadcasting Operation Systems

Broadcasting operation systems are the technical core of the audio production of every modern radio station. The systems include function modules for capturing, archiving, scheduling and broadcasting on-air. The minimisation of the complexity of broadcasting operation systems allows broadcasters to launch new programme formats with less effort.

Audio takes are captured data reduced in the MPEG-1 audio format. In order to avoid reductions of audio quality, in cases of encoder cascading, the archived data rate should be chosen with the highest value (i.e. 384 kbit/s). During the archiving process, additional text information, like title, track and artist, is also captured. This information can easily be used as ancillary data for PAD information. The broadcasting operation systems work in two different modes: full automatic or with moderation in a semi-assistant mode.

Editors arrange the stored audio information in daily play lists. During the runtime of the play lists, the stored audio takes will be replayed by MPEG PC cards (see also Figure 5.2).

5.2.2   Editorial Systems

The editorial tools based on computer systems and servers, provide large electronic information resources for the daily work of broadcasters. On completion of contributions, those ready for post-production are passed to data services. Suitable tools, like macro languages from desktop publishing systems (DTB), provide the content for the use in HTML files or pictograms.

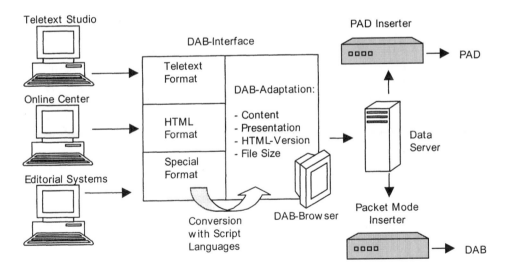

Figure 5.2   Editorial systems and DAB data interfaces

In this context helpful script languages such as PHP3 (Personal Home Page), PERL and JAVA are employed. Figure 5.2 shows an arrangement of an editorial system with the necessary DAB interfaces.

The resulting data files are transported via local area networks to the connected data servers. Internet template tools are very popular today in on-line service centres. With these text template systems, the service design can be made with pre-defined text fields which have to be completed by the editors. This process allows to individually generate content for specific applications such as Internet, DAB Broadcast Web Site or DAB Dynamic Label headlines.

5.2.2.1   On-line Service Centre

The optimal information source to provide DAB data services is on-line service centres. With the success of the World-Wide Web and the increasing numbers of connected users, many broadcasters have established on-line service centres on the Internet. The on-line editors are experienced and trained in the use of modern authoring tools.

An example of on-line information, which is transported in a PAD channel, is the DAB news service of "B5aktuell" in Bavaria. Actual news and background information are extracted from the on-line content in a multimedia database. An additionally developed software module recognises the HTML header tags, which include actual headlines, and generates a news ticker for a Dynamic Label service.

5.2.2.2   Teletext

The teletext service, which was introduced during the 1980s in European countries, is one of most popular ancillary services of broadcasting systems. Owing to the increasing numbers of teletext users, broadcasters established their own teletext offices in-house. New TV stations are merging teletext and on-line centres into one common division.

Teletext is primarily the responsibility of TV editors, but if radio managers are convinced about the vision of a "mobile teletext" over DAB, the teletext system can be used as an easy and economic data source. Such a service has been running in Bavaria, Germany, for several years. The DAB network provider uses a PC-based TV-receiver card with an integrated teletext decoder to provide a packet mode data service. The received teletext files of the *Bayerischer Rundfunk* are composed with the script language PERL into a new DAB Broadcast Web Site service. This service provides information on traffic, news, sport, business and the flight schedules of Bavarian airports.

5.3   Needs for New Infrastructure

The previous section described adaptations to existing infrastructures. In most cases the requirements of programme editors can be satisfied. Sometimes such approaches lead only to spot solutions, which are limited for future expansion. In this case completely new planning is necessary. The present section shows current developments for a structured management of text, service information and multimedia content.

5.3.1   *Management of Text and Service Data*

With the introduction of ancillary data within the Radio Data System (RDS) for FM radio, Astra Digital Radio (ADR) and DAB in Germany, new needs for a programme-oriented administration of ancillary data resulted at German broadcasters. One of the main objectives was the provision of uniform data to the data inserter. The result of this requirement was the concept of the Ancillary Data Manager and the in-house data protocol called "Funkhaustelegramm".

5.3.1.1   Ancillary Data Manager

The PC-based software Ancillary Data Manager was developed for the accumulation and distribution of ancillary data for a single radio service. The system concentrates, controls and provides information collected from sources like broadcasting operation systems, editorial modules and studio contacts. The concept described appears under different product names in Germany (see Figure 5.3).

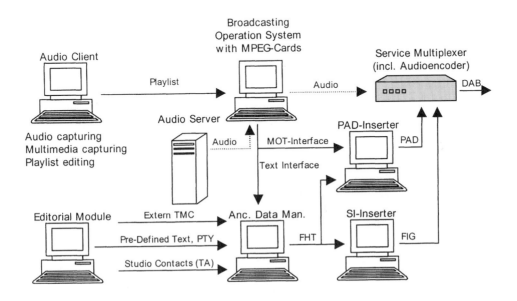

Figure 5.3   DAB studio infrastructure

Module Broadcasting Operation System

A text file interface to broadcasting operation systems allows synchronised adaptation of text information to the current audio item. The file interface stores information (i.e. title, artist, ordering information) after modification or new starts of audio takes. The contents of the file will be deleted at the end of items.

Module Editorial System

This module stores in a database pre-defined scheduled information like dynamic Programme Type (PTY), text masks, event texts or moderators. The text masks allow a more attractive presentation of titles, artists, times, moderators and magazine names.

Radiotext example:
"[MODERATOR] presenting at [TIME] the title: [TITLE] performed by [ARTIST]".

Module Relay Interface
A PC-relay card detects studio switching contacts, such as a fader (signal level attenuator) contact in a traffic studio, for the generation of a traffic announcement (TA). The system supports 10 different announcement types (e.g. traffic, sport, news, transport, etc.). The relay interface also recognises the music or speech status, for generation of the M/S flag, detecting the mixing console fader.

External Data

An interface for pre-produced information allows the feeding of formatted "Funkhaustelegramme" (see section 5.3.1.2), that is TMC messages or DGPS information.

Control unit

The central system part automatically schedules and formats the outputting data. Adjustment of different priorities and broadcast cycles of the input modules described above allows an optimal presentation mix.

For different applications, such as differences in the text lengths of DAB text (128 characters) or RDS text (64 characters), the configuration of different outputs is possible.

Basic adjustments can be made to the configuration of static information, namely Programme Source (PS), Programme Identification (PI) or Traffic Programme (TP). Table 5.3 shows the information types of the system.

The ancillary data manager formats the output data according to the RDS transmission protocol, see (ARD, 1994) or UECP (Universal Encoder Communication Protocol).

Table 5.3 Ancillary Data Manager information types

Type	Information
CT	Clock/Time
DI	Decoder Identification
DGPS	Differential Global Positioning System
MS	Music/Speech
ODA	Open Data Application
PI	Programme Identification
PS	Programme Source
PTY	Programme Type
RT	Radiotext (64 and 128 characters)
TA	Traffic Announcement and 10 add. types
TDC	Transparent Data Channel
TMC	Traffic Message Channel
TP	Traffic Programme and 10 add. types
TTA	Title – Track – Artist (only ADR)

5.3.1.2   In-house Protocol

The specially defined in-house protocol "*Funkhaustelegramm*" is a recommendation of the heads of the ARD (German public broadcasters) radio production for the distribution and transmission of ancillary data in broadcasting houses. One of the main objectives of the protocol definition was a uniform format for providing ancillary data to the data inserters for RDS, ADR and DAB (PAD and SI). The "*Funkhaustelegramm*" (see Figure 5.4) has a variable length and consists of three fragments: data head, data content and check sum. The data head contains a sync word, length, information type and source and destination addresses. The recommendation considers all control, service and text information types of RDS, ADR and DAB. The source and destination addresses allow individual addressing of a certain data inserter and recognition of the studio. The data are transmitted to the data inserter via local area networks or serial data lines.

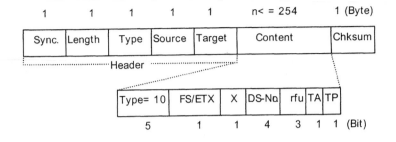

Figure 5.4   Structure of the "Funkhaustelegramm" (i.e. traffic announcement, TA)

5.3.2   Multimedia Editorial Tools

5.3.2.1   Multimedia Interface for Broadcasting Operation Systems

The broadcasting operation system (see Figure 5.3) allows a synchronised broadcast of audio and associated information. The capturing and administration of multimedia content requires an expansion of the existing audio database model. During the hard-disk recording of the audio takes, any multimedia content (i.e. CD covers, pictures of artists or advertising pictures) can be stored in the database and combined with the associated audio parts. The IRT recommendation "Multimedia Object Transfer (MOT) Interface for Broadcasting Operation Systems" (see Table 5.4) has the same approach for multimedia content as the "Funkhaustelegramm" for text information. The MOT interface ensures a synchronised handling of multimedia information in the data inserters or web servers. Therefore, the broadcasting operation systems have to create a dBaseIII-compatible file, containing the

Table 5.4   MOT interface for broadcasting operation systems

Name	Type	Length	Description
DATE	Date	–	Date
TIME	Char	8	Start time
TITLE	Char	36	Title
ARTIST	Char	24	Artist
H	Num H	6	Length in hours
MIN	Num H	6	Length in minutes
SEC	Num H	6	Length in seconds
PARFILE01	Char	60	Path and file name of the parameter file
MOTFILE01	Char	60	Path and file name of the MOT object file
...	...	...	...
PARFILE10	Char	60	Path and file name of the parameter file
MOTFILE10	Char	60	Path and file name of the MOT object file

MOT objects, which is read by the data inserters. The interface guarantees a fixed connection to the programme schedule. Every change in the play list is recognised by the data inserters.

The MOT interface is designed to store information ahead, over n titles or n minutes. It is the decision of the editor responsible to configure the associated information ahead of/or contemporaneously to the audio item. In the first case, the trigger time of the object, included in the MOT header, is set according to the play list time. In the second case, the trigger time is set to "now". This means that the MOT object will appear in the receiver immediately after transmission. Furthermore the MOT interface contains directory information for the objects and parameter files. This information indicates the archive file location. The parameter files contain pre-encoded MOT information (e.g. file size, object type, service manner). During the insertion process the PAD inserter requires all information.

5.3.2.2   Authoring Tool "DAB Slide Show"

The "Slide Show" application consists of sequences of JPEG or PNG pictures. A simple form of this service is the presentation of CD covers or web camera pictures. In the editorial tasks there is the desire to mix single content sources to provide an attractive service. Based on this desire, the German broadcaster "Westdeutscher Rundfunk" developed some software named "PAD-Projector". This allows the configuration of programme categories, such as weather, traffic, music or advertising. Every category consists of a picture and a text layer. The picture layer allows easy recognition of the topic by the use of pictograms. The text layer consists of actual information. Both layers are stored in a JPEG file and transmitted to the data inserters or web servers. The settings allow the configuration of default pictures (in the case of no available actual text content), the JPEG compression rate, the duration and the order of the slides.

5.3.2.3   Encoded Traffic Information (TMC, TPEG)

The objective of encoded traffic information is to avoid disturbances of formatted radio. Monotonous announcements and a large stock of messages have a negative influence on modern programme making with fixed speech/music portions and uniform programme flow. The objective of an exclusive coverage with encoded traffic information is, today, due to less market penetration with suitable receivers far away. Nevertheless it is important now for broadcasters to build up the necessary infrastructures to compete in future with other content traffic information bearers, like mobile telephone systems.

The RDS TMC services generated by traffic information systems currently provide a regular service in European countries. Car navigation systems are using RDS TMC for free updating of traffic status. Therefore, for identical coverage of the existing RDS market a TMC provision in DAB is essential. TMC was designed for transport in a narrow-band RDS data channel and because of that the standard is restricted for future expansion. Founded by the European Broadcasting Union (EBU), an independent expert group TPEG (Transport Protocol Experts Group) is now developing a standard for the transmission of traffic and transport information within DAB, DVB and the Internet.

5.3.2.4   Multimedia Database

A multimedia database is under development at the IRT for central contribution and distribution of multimedia content. The database is designed as an SQL database supported by PHP3 and JAVA scripts running on LINUX or WindowsNT servers. The input scripts are adapted for single editorial applications and convert the arriving documents (DTP document, email) into a uniform database format. The output scripts can be optimised to the bearers (Internet, DAB, WAP, DVB) and provide applications like HTML, "Slide Shows" or WML (Wireless Mark-up Language). The authoring tools are run easily with Internet browsers. The aim is to develop a growing multimedia toolbox for new broadcasting media with tools from the Internet market.

5.3.3   Data Inserter

5.3.3.1   SI Inserter

Service Information (SI) is transported in the Fast Information Channel (FIC) and includes static information (i.e. Programme Source PS) and dynamic information (i.e. announcements, TMC).

SI is configured by the control software of the service or ensemble multiplexer. Dynamic SI is inserted with an SI inserter (see Figure 5.5) into the DAB data bit stream. Therefore the service multiplexer needs an additional input interface. The ancillary data manager of each programme (see section 5.3.1.1) and the protocol "Funkhaustelegramm" (see 5.3.1.2) are operating as information sources. Essential dynamic SI is TMC, DAB announcements and dynamic programme type PTY. Additional basic configurations allow the signalling of service provider ID, service ID and sub-channel ID. Furthermore, a cluster ID bundles several audio services to support announcements.

The received "Funkhaustelegramm" is stacked into input buffers and prioritised in sending order. The assignment to the programme is made either with the source information of the delivered data or hard-wired with the connected serial interfaces. The SI is encoded as Fast Information Groups (FIG) and transported in STI-D format over a serial RS232 interface to the connected service or ensemble multiplexer. A cyclic repetition of the FIG data, according to the DAB Guidelines of Implementation, ensures error-free and quick

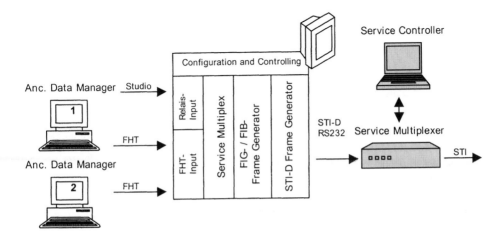

Figure 5.5   Service information inserter

recognition at the receiver. For the future, the implementation of backward functionality over the STI-C channel is planned. This allows transportation of reconfiguration information from the network providers to the studio of the content provider.

5.3.3.2 PAD Inserter

The ISO/MPEG audio standard covers the transport of ancillary data within the audio frame. These data are fed into the audio encoder via a serial RS232 interface (according to the IRT Recommendation "Data Interface for ISO/MPEG Audio Codecs and Ancillary Data Devices").

The PAD inserter (see Figure 5.6) pre-formats the ancillary data and works as an interface between the editorial environment and the DAB technology. The DAB standard provides several PAD information types (see section 5.3.2.1). In most DAB projects in Europe the applications Dynamic Label, MOT Broadcast Web Site and MOT Slide Show are used. The figure illustrates the functionality of a PC-based PAD inserter system (IRT concept) with input interfaces for text and MOT objects, provided over serial interfaces or local area networks.

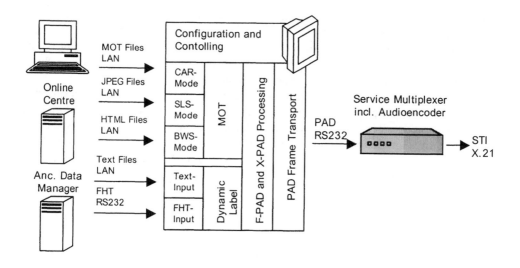

Figure 5.6   PAD inserter

The modes MOT-BWS and MOT-SLS are basic modes for the transmission of interactive or automatic services. In the respective modes there are a selection of homepage files for fast transport within the data carousel or configuration of a gap time for the Slide Show sequences possible. The mode MOT-CAR allows the remote control of the PAD inserter with broadcasting operation systems. In this mode, the data inserter is directly connected to the schedule of multimedia objects. Furthermore, an additional PAD scheduler allows the provision of a daily service disposition, for instance:

10:00 AM – 11:00 AM Service: MOT-BWS.

5.3.3.3   Packet Mode Inserter

Two approaches to packet mode inserter systems, depending of the operating location, are possible. In the first one, the packet mode inserter operates at the location of a service provider. For that purpose, similar interfaces as at the PAD inserter system have to be implemented. In the second approach, the packet mode systems operate at the network provider's location. This is necessary if public and commercial contents are provided from one of the network provider's server. This scenario, currently operating in Germany, needs additional requirements such as the editorial interfaces described above. The functionality described below (see Figure 5.7) is based on the concept of the "Multimedia Data Server" developed by Frauenhofer Gesellschaft (FhG).

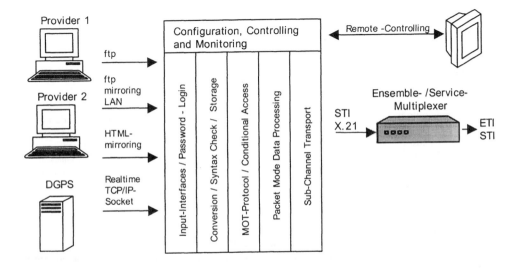

Figure 5.7   Packet mode inserter (based on the Multimedia Data Server developed by FhG)

The central system components comprise the reception, verification, conversion, storing and sending of packet mode data. The reception of content from different providers is handled over modems or Internet connections. The individual login to the provider's directories is ensured with the help of login scripts and password detection. After reception of a file, the HTML syntax and version will be verified and if necessary modified. The valid data are prioritised within the data carousel, according to the ordered packet mode data rate, and sent as MOT objects within a packet mode channel over an X.21 STI-D interface to the ensemble multiplexer. During the Bavarian DAB project data scrambling (VIACCES method) and the transmission of time-critical D-GPS correction data were also successfully demonstrated.

5.4   Compatibility with RDS

Some special features such as data services or other ancillary information are already being used in conventional radio services. One very well-known service is the Radio Data System

RDS (EN 50067). It is important to consider aspects of the implementation of the RDS and DAB systems which are directly related to each other to ensure as much as possible a consistent and compatible approach. In other words, it is important to examine how the various features of both systems can best be implemented both by broadcasters and in a receiver to allow a user to benefit from both systems. The listener must be able to access services and features in a similar manner regardless of whether tuned to either FM/VHF or DAB.

5.4.1   DAB SId Versus RDS PI Code

Both Programme Identification (PI) codes in RDS and Service Identifier (SId) codes in DAB use a 16-bit code to identify audio services. Although the code structure may appear identical in RDS and DAB, usage of the code in terms of allocation and broadcaster and receiver implementation differs in the two systems as illustrated in Table 5.5.

Table 5.5   Differences between RDS PI and DAB SId

RDS PI	DAB SId
The second nibble of the 16-bit PI code represents the coverage area of the RDS service (Local, International, National, Supra-regional, or one of 12 Regions). This imposes constraints upon free PI code allocation. Services with PI codes that are identical in the first, third, and fourth nibbles are implicitly regional equivalent services, i.e. they are "soft-linked".	The second nibble of the 16-bit SId code has no special significance, and there is no "soft linkage" implied between any two SId codes. "Soft linkage" is explicitly given by means of the linkage feature (FIG 0/6) by using common Linkage Set Numbers (LSNs).
Regional RDS services often use separate codes when originating their own programmes, but transmit a common PI code during shared programming. Broadcasters who vary PI codes in this manner are unable to provide identical codes on DAB and RDS even though services will at times be carrying identical audio.	For regional services SId codes must remain unchanged and be allocated unique codes, even if RDS services change their second nibble during common programming (in DAB SId codes must remain invariant). In this case, as a consequence, in any region the PI code and SId code will be different even though programmes are simulcast. A DAB service is explicitly linked to regional equivalent services using linkage.

If RDS and DAB services carry the same programme at all times, common PI and SId codes may be allocated (SId = PI) and the SId code allocation will be constrained by the same rules as for RDS codes. The receiver can treat a common code as an implicit hard link. Service following between DAB and RDS can be done in this case without using the linkage mechanism. A receiver may use services with identical codes as alternative sources of the same programme. If at any time it is likely that different programming is carried on RDS and DAB, then different PI and SId codes must be allocated. In this case following between RDS and DAB can only be done by making use of the linkage feature.

To follow between RDS and DAB services without using linkage, the following may be noted: a receiver, on seeing that the RDS PI code has a second nibble of either "2" or "3", will know that all regions are carrying a common programme. Hence a switch is allowable to any RDS or DAB service whose PI or SId code is identical in the first, third and fourth nibble and where the generic (EG) flag is set.

5.4.2   Programme Type Codes

The Programme Type (PTy) feature in DAB has been designed to give improved functionality compared to the Programme Type code (PTY) in RDS. Consequently all related features or possibilities in DAB are not available in RDS.

Table 5.6 illustrates the differences between PTY (RDS) and PTy (DAB). In DAB it is possible to transmit PTy codes to represent both the static (service format) and dynamic (actual on-air programme item type) simultaneously. For both static and dynamic codes, up to two codes may be used. For each of the static and dynamic codes, at least one is chosen from the same international code range as used for RDS. The second, if used, may be chosen from this same set, or from the additional range of 32 coarse codes.

Table 5.6   Differences between PTy in DAB and PTY in RDS

PTy in DAB	PTY in RDS
Besides of the basic set of 30 internationally agreed Pty codes a further set of 32 coarse codes and up to 256 PTy fine codes are available to categorise the programme content.	A list of 32 internationally agreed PTY codes is available to categorise the programme content.
Up to eight PTy codes can be allocated at the same time to a programme service.	A single PTY code only can be allocated to a programme service.
PTy codes can be signalled for the selected services and for other services via the PTy and other ensembles PTy feature.	PTY codes can be signalled for the selected service as well as for another service via EON (Enhanced Other Network).
PTy fine codes allow finer sub-divisions of some of the PTy coarse codes and can be used to watch for more specific programme types (e.g. "Tennis" instead of "Sport"). To preserve as much compatibility as possible, the DAB "fine" code should match the RDS PTYN description as much as possible.	There are no PTY fine codes in RDS but the Programme Type Name (PTYN) allows a broadcaster to transmit a short eight-character description to describe the programme content. The characters are transmitted over the air, and cannot be used for PTY Search or Watch functions.
PTy language allows PTy selection based on language criteria (e.g. "News" in "Swedish").	PTY language is not supported in RDS.
PTy downloading allows new codes to be defined "over the air". For PTys which are not agreed internationally, an accurate description can be provided. It also allows the description of PTy codes to be translated into any desired language.	PTY downloading is not supported in RDS.
PTy preview signals which PTy codes are likely to be used in the next one or two hours. This assures the listener whether a WATCH request is likely to succeed or not.	PTY preview is not supported in RDS.

5.4.3   Announcements in DAB versus TA in RDS

Table 5.7 illustrates the differences between the Announcement feature in DAB and Traffic Announcement (TA) in RDS.

Table 5.7   Comparison between announcements in DAB and TA in RDS

DAB	RDS
16 announcement categories can be coded, of which 11 are currently defined. Announcement types are treated in a different way as vectored PTys in WATCH mode, whereas dynamic PTys are scheduled, announcements are unscheduled and used for short vectored interrupts.	In RDS only one announcement type (TA) is defined. The dynamic use of PTy codes 01 (News) and 16 (Weather) is often used to provide similar functionality to announcements in DAB.
Announcements are based on the cluster concept. They can be targeted to specific regions and feature a "New" flag, which is valuable in the case of a cyclic announcement channel. In the receiver announcements can be filtered according to the intended region.	TAs (or dynamic PTYs) with cluster operation is not supported in RDS. A "New" flag is not supported in RDS. Regional filtering of TAs is not supported in RDS though TAs from regional stations will inherently be regional.

In RDS some dynamic PTY codes are used as fallbacks for DAB announcement types. Table 5.8 illustrates the coding of programme items intended to cause vectoring to a short audio message or "announcement".

Table 5.8   Coding of programme items intended to cause vectored interrupts

Description	Coded in DAB as:	Coded in RDS as:
Alarm	Anno flag b0	Dynamic PTY 31
Road Traffic	Anno flag b1	TP/TA
Public Transport	Anno flag b2	TP/TA
Warning/Service	Anno flag b3	Not available
Alarm Test	Not available	Dynamic PTY 30
News	Anno flag b4	Dynamic PTY 01
Weather	Anno flag b5	Dynamic PTY 16
Event	Anno flag b6	Not available
Special Event	Anno flag b7	Not available
Programme Information	Anno flag b8	Not available
Sport report	Anno flag b9	Dynamic PTY 04
Financial report	Anno flag b10	Dynamic PTY 17
Reserved for future announcement types	Anno flags b11–b15	Not available

Some broadcasters may choose to use the announcement categories "News flash" and "Area weather flash" instead of, or in conjunction with, programme types "News" and "Weather" respectively. As a consequence, in DAB, PTy codes 01 and 16 may be used for longer as well as shorter programme items, and in addition may be combined with a language filter code.

5.4.4   Dynamic Label in DAB Versus Radiotext in RDS

Both RDS and DAB have the ability to carry text messages from the broadcaster to the listener's receiver. Although designed for the same purpose, Radiotext in RDS and Dynamic Label in DAB have a number of differences, see Table 5.9. In particular, control code use is different. There is a possibility to harmonise the usage to allow both broadcasters and receiver designers to implement the same text handler routines and achieve acceptable results and presentation on both systems.

Displays, which rely upon scrolling the text to show the complete message, should apply intelligent routines to present the message in the best possible way. Receivers could remove unnecessary spaces inserted by the broadcaster as "padding", before display, to prevent scrolling through strings of blank spaces. If the resultant text length is shorter than, or equal to, the number of characters in the display, the display could show the message statically, without scrolling. Although the primary responsibility for presentation rests with the receiver manufacturer, broadcasters need to have a basic knowledge about the way Dynamic Label and Radiotext messages are being treated in receivers.

It is therefore up to the broadcaster to ensure that the data stream transmitted does not cause undesirable effects on older generations of RDS receivers. The receiver designer is responsible for ensuring the optimum presentation of the text, appropriate to the size and type of display chosen.

Table 5.9   Differences between Radiotext in RDS and Dynamic Label in DAB

RDS Radiotext	DAB Dynamic Label
Supports a 64-character message.	Supports a 128-character message.
8-character displays on some older receivers.	Wide range of displays from 1x16 or 2x16 characters, LCD screens and PC monitors
Multiline displays should use normal word-processing techniques to provide word wrapping to show the message completely.	Because of the wide range of display types possible, it is up to the receiver manufacturer to develop software to optimise the presentation for the particular display chosen. Multiline displays will need to accommodate more than 128 characters to allow for word wrapping to show the message completely.

5.4.4.1   Control Codes

A number of control characters, which the broadcaster may choose to insert in the text string, have been defined to create a particular effect on certain displays. As some of these control codes were not defined in the original RDS specification, some older RDS receivers may not be able to respond to these codes in the correct way. Most first-generation RDS receivers will substitute a space for these control codes.

a)   Control code "0B"

Control code 0B is used to indicate the end of the portion of text the broadcaster has designated as the "headline". The headline is a portion of the text meant for display in a fixed format containing information that does not require frequent updating and that is safe for viewing by driving motorists. It is physically sent as the first portion of the message and comprises all characters up to, but not including, control code 0B. If a headline is present, the "body" is the part of the text excluding the headline and control code 0B. In compiling messages, the broadcaster ensures that the headline portion must be capable of standing alone, and that the complete message must make sense. The body may not necessarily make sense on its own. Receivers may choose to display:

- only the headline portion of the message, or
- the complete message with headline portion enhanced or differentiated, or
- the complete message without regard to the headline status,
- but *not* the body of the message alone.

If 0B occurs in the current message, it defines the headline, which comes into effect immediately. If no 0B occurs in the current message, there is no headline in effect. If 0B occurs in position "0" no headline is defined. The headline starts in position "0" and runs until the position preceding code 0B. Its effective length must not exceed 32 displayable characters. The occurrence of code 0B and hence of the headline is optional.

The content of the headline should not be changed more often than is compatible with road safety requirements. This does not impose restrictions upon the content of the body of the message which can be updated more frequently, nor on the repetition rate of the headline.

In the usual case where the broadcaster intends a space between the headline and body parts of the message, a space character must be explicitly transmitted.

- **In the case of DAB receivers and compatible RDS receivers:** When no headline is available, the receiver designer must decide upon the appropriate actions. Code 0B does not occupy a space in the displayed message. Code 0B, when ignored by the receiver (because no headline display mode is supported), does not represent a space. As the body text may not make sense on its own, it must not therefore be presented in its own right.

- **In the case of first-generation RDS Radiotext receivers:** As code 0B was not defined in the original RDS specification, some older RDS receivers may not be able to distinguish between the headline and body portions of the message. They may display headline and body as a single text string. Most receivers will substitute a space for 0B. Broadcasters including code 0B in their Radiotext messages must be aware of this behaviour.

b)   Control code "0A": preferred line break

Control code 0A is used to indicate positions in the message where the author of the text would prefer a line break to occur to structure the text in a particular way.

When control code 0A is used in the headline portion of the text, it indicates the position at which a receiver, using a two-line, 16 characters per line display, should preferably break the headline. In case of a larger format display, the receiver may choose to ignore it.

If code 0A occurs in the headline, then the sum of displayable characters must be less than or equal to 31 (because a receiver equipped with a 1x32 character display format will insert a space character in place of the control code).

Code 0A must not occur more than once in the headline portion of the message, and may not be used if a 1F (preferred word break) is also used in the headline portion.

- **In the case of DAB receivers and compatible RDS receivers:** If a receiver uses a one-line, 32 character display to present the headline portion of the text, code 0A within the headline, if used, is replaced by a space in the display and not used to create a line break. If a receiver uses a 2x16 character display to present the headline portion of the text, the character string before code 0A is displayed in the first 16-character block; 0A codes are transmitted to aid a receiver to structure the display of the message optimally. Code 0A is replaced by a space in the display when not used to create a line break.

- **In the case of early RDS RT receivers:** Since code 0A was not defined in the original RDS standard, some older RDS receivers may not be able to use it to create a line break if desired. Most of these receivers will substitute a space for code 0A. Broadcasters including 0A in their Radiotext messages should be prepared to specifically transmit a space character in conjunction with code 0A to prevent unintentionally joining together the two words either side of the 0A code on certain displays.

c)   Control code "1F": preferred word break (soft hyphen)

Control code 1F indicates the position(s) in long words where a receiver should preferably break a word between multiline (non-scrolling) display lines if there is a need to do so.

For example: If control code 1F is used in the headline portion of the text, it indicates the position at which a receiver using a two-line, 16 characters per line display should preferably break the long word. If code 1F occurs in the headline, it divides the headline in such a way that neither sub-string contains more than 16 displayable characters.

Code 1F must not occur more than once in the headline portion of the message, and may not be used if a code 0A (preferred line break) is used in the headline portion.

- **In the case of DAB receivers and compatible RDS receivers:** Code 1F, when used by a receiver as a word break, is replaced by a hyphen followed by a line break, otherwise it is ignored and does not represent a space.
- **In the case of early RDS RT receivers:** Devices produced to the original RDS specification (EN 50067) are unable to use it to create a word break. Most of these receivers will substitute a space for code 1F, so that words will be unintentionally split at the point of the 1F. For this reason, broadcasters may decide that use of code 1F is inappropriate for Radiotext.

d)   Control code "0D": end-of-text

Control code 0D is used in RDS to indicate the end of the current message. It is used to help RDS receivers identify the actual length of the current message if it is less than the maximum 64 characters possible. DAB receivers do not use this control code.

Code 0D can appear only once in a message. Positions after code 0D should be disregarded and should not be transmitted or filled by spaces.

- **In the case of DAB receivers and compatible RDS receivers:** Code 0D and any characters or codes in all the following positions are ignored.
- **In the case of first-generation RDS RT receivers:** Since control code 0D was not defined in the original RDS specification, some older RDS receivers may be unable to use it. Most of these receivers will substitute a space for code OD, others may delete it from the string.

5.4.5   Cross-referencing DAB Services from RDS

It is likely to be a number of years before DAB coverage equals that of FM radio. During this period when DAB will be unavailable in all areas or for all services, it is assumed that receivers, especially vehicle receivers, will be able to tune to both DAB and FM transmissions.

Since DAB provides superior quality reception to FM, it should be the preferred choice for listening where available. Receivers may switch across from FM to DAB as soon as possible after entering a DAB served area if the DAB tuner knows in which ensemble, and on what frequency, the required service may be found.

The DAB cross-reference feature defines how to signal DAB frequencies within the RDS format to achieve a fast and effective way for a combined RDS/DAB receiver to get access to alternative programme sources in DAB transmissions.

5.4.5.1   ETSI Standard EN 301700

The European standard (EN 301700) describes the way that DAB service information can be provided from RDS by using an Open Data Application (RDS-ODA; the RDS Forum allocated the number AID = "147") allowing a receiver to find an equivalent DAB service. The ODA allows a service provider to signal not only frequency and transmission mode information about DAB ensembles but also linkage information and ensemble information about DAB services. This information can be used by a receiver to perform service following from RDS to DAB. The ODA uses an RDS type A group so that there are 37 bits available for coding. One bit is used for the *E/S flag* to differentiate between data for the ensemble table and data for the service table. The remaining 36 bits are used to express ensembles in terms of frequencies and modes, or services in terms of ensembles and service attributes (PTy, LSN, etc.).

This information will enable the tuner

- to find the location of DAB ensembles
- to find where the current RDS service is on DAB
- to find a list of other DAB services
- to know some of the attributes (PTy, LSN) of DAB services.

5.4.5.2   Relationships between DAB Services, Ensembles and Frequencies

In RDS, there is a single relationship: a service is carried on one or a number of *frequencies*. The RDS system allows a service provider to send all the frequencies the service is available on to RDS receivers. The RDS receiver can build up a list of alternate frequencies (AF) which allows (when mobile) to find the best frequency for that service. The RDS system also allows other service information to be provided, and, through the Enhanced Other Network (EON) feature, this can be extended to other RDS services.

In DAB the situation is different in that services are carried within ensembles. A given service may be carried in one or more ensembles and each ensemble may be carried on one or more frequencies. A DAB receiver is therefore required to maintain two tables, one listing ensembles and attributes for each service (*Service Table*), and a second, which lists frequencies and mode for each ensemble (*Ensemble Table*).

The requirement for cross-referencing from RDS is to deliver data to the DAB receiver to allow these tables to be constructed.

5.4.5.3   Ensemble Table

The ensemble table contains the basic information required for service following:

- the Ensemble Identifier (EId),
- the DAB transmission mode
- the centre frequency of the DAB ensemble.

Reception of the ensemble table does not mean that the tuned RDS service is carried on that ensemble. The purpose of the ensemble table is to allow the receiver to build up information about DAB ensembles. To locate an equivalent DAB service for service following requires that the receiver tune to each ensemble and inspect the service linking information. This task can be simplified by using the service table (see section 5.4.5.4).

Table 5.10 illustrates the ensemble table format. The EId being referenced occupies the last 16 bits in block 4. The preceding 16 bits in block 3 are used to signal the frequency code. In block 2, 2 bits are used to indicate the transmission mode.

NOTE: In (EN 300401) the centre frequency of an ensemble is coded as an unsigned binary number in a 19 bit field, allowing a range from 16 kHz to 8388592 kHz. As in this ODA only 18 bits are available, the frequency range will be limited from 16 kHz to 4194288 kHz.

Table 5.10   Ensemble table format

PI code (16 bits in block 1)	Mode field (2 bits in block 2)	Frequency field (16 bits in block 3 + 2 bits in block 2)	EId field (16 bits in block 4)
(PI code of tuned RDS station)	00 = unspecified 01 = mode I 10 = mode II or III 11 = mode IV	centre frequency of the DAB ensemble in the range 16 kHz – 4194288 kHz	EId of the DAB ensemble to which transmission mode and frequency apply.

A transmission of this RDS group is required for each EId and for each different frequency. Once all frequencies have been broadcast for a particular EId, another EId with its list of frequencies may then be broadcast. Every group is complete in its own right, and data are usable directly, without reference to any other group.

5.4.5.4 Service Table

The service table provides additional information about services available on DAB. It allows a receiver to get information via the RDS-ODA rather than having to examine each DAB ensemble for FIC information. Variant 0 allows the receiver to build up a list of ensemble identifiers that a service is available on, and variant 1 allows the linkage information for the DAB service to be stored. The service provider will signal every ensemble that a service is carried in before signalling the other attributes for that service. Once all the information is signalled for one service, the information for the next service may be signalled, and so on.

Table 5.11 shows the service table format. The SId being referenced occupies the last 16 bits in block 4. The preceding 16 bits in block 3 are used to transmit the list of Ids and all other required attributes of that service. The use of these 16 bits is indicated by the variant code that occupies the last 4 bits in block 2.

Table 5.11   Service table format

PI Code (16 bits in block 1)	Variant (4 bits in block 2)	Information Block (16 bits in block 3)	SId (16 bits in block 4)
(PI code of tuned RDS service)	(0–15)	variant 0: Ensemble information: EId variant 1: Linkage information: Linkage Actuator (LA), Soft or Hard link International Linkage Set (ILS), Linkage Set number (LSN), Rfu	SId of the DAB service to which the information block data applies.

A transmission of this RDS group is broadcast for each SId for each different EId. Once all EIds have been broadcast (using variant 0) other variants are transmitted to describe the other attributes for that particular SId. Other services may then be broadcast, each with its own list of EIds and service attributes. Every group is complete in its own right, and data are usable directly, without reference to any other RDS group. The transmission of the service table information is optional but when it is transmitted then both variant 0 and variant 1 information should be broadcast.

5.4.5.5 Operational Requirements

In order for the (RDS) receiver to know which application decoder to use and which group type is being used for carrying the ODA (Open Data Applications) data, the service provider must transmit two type 3A groups per minute.

If a service provider has many services carried on DAB in many ensembles up to two groups per second may be required to transmit all the application data. The ensemble and service table carousel may be interleaved but data will be transmitted in sequence within each carousel. All the ODA data will be transmitted within 2 minutes.

As already explained in section 5.4.5.3, application data from the ensemble table are transmitted such that all data for one ensemble are broadcast before data for the next ensemble are broadcast. As explained in section 5.4.5.4, application data from the service table are transmitted such that all data for one service are broadcast before data for the next service are broadcast. All ensemble information (carried in variant 0) relating to one service is broadcast before other data (carried in other variants) relating to that service are broadcast.

For service following it is recommended that the receiver use the linkage information provided via RDS for the RDS service and via DAB for the DAB service to link between equivalent services. This means that basically only the ensemble table needs to be transmitted. However, single front-end receivers have to use the linkage information provided via RDS for both the RDS and the DAB service, and therefore both the ensemble table and the service table should be transmitted. The linkage information is provided either explicitly by RDS type 14A groups and DAB FIG type 0 extension 6 (see section 5.5.2) or implicitly when the RDS PI code and DAB SId are identical (see section 5.4.1 above).

5.5   Audio Service Aspects

5.5.1   Loudness Differences

A mostly occurring technical and acoustical problem during audio production is the loudness differences within or between radio programmes. Basically this problem contains three aspects:

Loudness differences within audio services
Audio engineers are recommended not to exceed an audio signal level of 100% (0 dBrel = +6 dBu). It depends on the experience of the audio engineer to recognise louder signals from the rhythm or retrace of the signal and to individually regulate their amplitude. Differences between speech and music items or between different audio sources (hard disk, CD jukebox) can be solved with suitable sound processors.

Loudness differences at DAB/FM switching
In limited DAB coverage areas a signal level of 100% corresponds at the analogue FM system to an FM deviation of 40 kHz. At digital signals the reference is fixed to –9 dBFS (full scale). The headroom of 9 dB is used as a control reserve for suddenly occurring signal peaks and should not be used for a maximum modulation. This recommendation is also valid for receiver implementation.

Loudness differences within DAB ensembles
This problem occurs mostly with different providers within a DAB multiplex. One reason for this can be different sound processing or a too low adjusted headroom of the analogue inputs of the audio coders. Here arrangements between the service providers are important for optimal presentation of the complete DAB system to the audience.

5.5.2   Audio Signal Level Alignment

Careful control of the audio level is very important in both the analogue and the digital environment. In the analogue environment, noise and distortion limit the capability of the system to accommodate signal with a wide dynamic range. In the digital environment, the system can be designed to accommodate such kind of signals, but it is essential to avoid hard "clipping" of programme signal peaks which occurs when the signal level is too high. However, there can be problems other than those directly concerned with dynamic range.

For example, different approaches to controlling levels may cause problems with the alignment of connections during the exchange of programmes between broadcasters.

The capability to handle signals with a wider dynamic range does not necessarily solve all the problems of programme level adjustment, see also (Gilchrist, 1998).

5.5.2.1   Programme Signal Metering and Control

In the past, broadcasters have used either the so-called Volume Unit (VU) meter or the Peak Programme Meter (PPM) to measure and set programme levels (IEC 268), (BS.645). The VU meter is a rectifier voltmeter with a rather long integration time (200 ms) which indicates approximately the average value of the voltage. It does not respond to short peaks in the programme. The PPM is a quasi-peak meter, with a rapid response to increasing level (the integration time is typically 10 or 5 ms respectively, depending from the definition of the standard), but the peak readings decay slowly. It gives a much better indication of the programme peaks than the VU meter, and the slow decay characteristic makes it easier for operators to observe the programme peaks.

There are a number of versions of the PPM, all having substantially the same ballistic characteristics, but with differences in the markings on the scale. Figure 5.8 displays the indications produced by various types of programme meters with the recommended test signals.

Most analogue systems do not introduce really "hard" clipping when overloaded, and the onset of distortion is usually gradual. This means that the occasional gentle clipping of short-duration signal peaks normally does not produce disturbing noise. Nevertheless, broadcasters will generally arrange for some "headroom" between the highest level indicated by the meter and the overload point of the transmitter, or provide a protective limiter, or similar device, to "catch" the highest signal peaks and reduce their level.

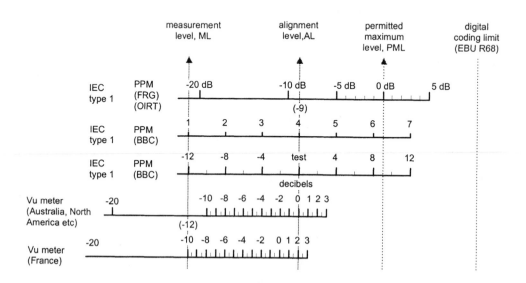

Figure 5.8   Programme level metering, according to (BS.645)

Electronic level meters (so-called true peak programme meters with a very short integration time < 0.1 ms) have been available for a number of years. They have the capability to capture and display accurately the level of even the briefest of signal peaks.

Any modification of programme peaks by the limiter may introduce distortion, or pumping, and the degree of annoyance depends upon the extent to which the peaks are modified. Some broadcasters, particularly those operating popular music stations, intentionally drive the limiters sufficiently hard to invoke the compression of all but the lowest signal peaks. This type of processing is used in order to produce a programme which sounds loud to the listener, possibly with the intention of sounding as loud as, or louder than, rival radio stations which may be competing for the attention of the listener.

5.5.2.2   Level Profile

The level profile of an audio transmission channel shows how the different levels of programme and test signals are located in the range between the upper system limit level (i.e. the clipping level in a digital domain, designed with 0 dBFS = 0 dB "Full Scale"), and the noise signal level.

Figure 5.9 shows the principal level profile of a transmission channel. For broadcasting systems (recording, distribution, emission) a set of audio test levels is defined by the ITU-R (BS.645) and the EBU (R68).

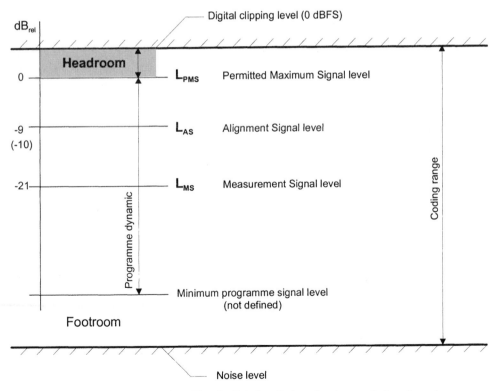

Figure 5.9   Principal level profile of a digital audio broadcast transmission channel

The "Permitted Maximum Signal level" L_{PMS} is recommended there as 9 dB below the digital clipping level (Full scale), and is intended to be related to the measurement of programme signal on quasi-peak meters that have an integration time of 10 ms, thereby ensuring that short transients are not clipped, because they will not fully use the provided 'headroom' of 9 dB between L_{PMS} and the system limit level. True peak reading meters will exceed this indication on some programme material, whereas VU meters will typically under-read this indication as they have a long integration time.

In broadcasting and some studio recording operations, where programme interchange compatibility is of primary importance, it is normal to work to international standard guidelines that define an "Alignment Signal level", L_{AS}, for alignment of the recording equipment or the transmission lines, respectively. ITU and EBU recommendations, among others, specify a digital alignment signal level of −18 dBFS, whereas SMPTE recommendations (RP200) specify a level of −20 dBFS (1 kHz tone, RMS measurement). Both are likely to be encountered in operational practice, and it is therefore important to indicate clearly which alignment level is adopted, in order to avoid subsequent confusion.

In addition, a lower signal level, the Measurement Signal level (L_{MS}) is defined at −21 dB below $L_{PMS,}$ for measurement purposes only (for instance measurement of the frequency response). This level has no significance for operational purposes.

5.5.2.3   Programme Levels and Level Alignment in Digital Broadcasting

Digital techniques offer the broadcaster the ability to handle audio signals with a considerably wider dynamic range than that accommodated by analogue techniques. Digital audio systems are characterised by noise and distortion levels which are determined principally by the parameters of the system (i.e. the resolution of the analogue-to-digital converters (ADCs) and any bit rate reduction techniques which may be employed) – and by "hard" clipping at the point where the amplitude of the analogue audio signal is sufficient to drive the ADC beyond the limit of its coding range.

With a sufficiently high digital resolution (i.e. 16 bits/sample for CD recording, but up to 24 bits/sample for an AES/EBU interface connection), connections for contribution, distribution and exchange can be implemented which provide a high signal-to-noise ratio and the capability to handle programme material with a wide dynamic range. The same is true for broadcast emission.

The application of bit rate reduction inevitably affects the noise levels, but practical digital radio services depend upon this for their existence, and today's advanced bit rate reduction techniques such as MPEG-1 or -2 Layer II used for DAB have little or no effect upon the perceived audio quality provided the audio signal is afforded at a sufficient bit rate (BS.1115).

The metering and control of audio signals should therefore conform to the recommendations of the EBU (R68-1992) and ITU-R (BS.645), as appropriate, in particular with regard to the provision of sufficient headroom, for instance 9 dB as recommended in (R68-1992), for transient signal peaks which exceed the permitted maximum signal level.

6

Collection and Distribution Networks

HANS-JÖRG NOWOTTNE and LOTHAR TÜMPFEL

6.1   General

6.1.1   Basic Requirements

Operation of the broadband DAB system requires the continuous collection of service component data, the formation of the respective ensemble multiplex data and their distribution to the transmitter sites of the single frequency network. This leads to a distributed, synchronised real-time system encompassing both a collection and a distribution network. These networks are connected by the ensemble multiplexer device which is situated in between.

Major differences between FM/RDS and DAB network architectures are outlined in Figure 6.1. In contrast to FM/RDS, DAB broadcasters, the service providers, send service component and additional data not directly to transmitter sites but on to the ensemble provider who operates the ensemble multiplexer. Tasks to be fulfilled by a DAB ensemble multiplexer are numerous owing to expanded functionality in DAB as provision for the dynamic change of the ensemble configuration, support of multicomponent services and enlarged on-air signalling compared to FM/RDS. With respect to signalling it should be emphasised that all service providers and the ensemble provider have to share the FIC, the common DAB signalling channel. Therefore provisions for dynamic signalling by service providers but also avoidance of interference or inconsistency of signalling data are important issues.

Digital Audio Broadcasting: Principles and Applications, edited by W. Hoeg and T. Lauterbach
©2001 John Wiley & Sons, Ltd.

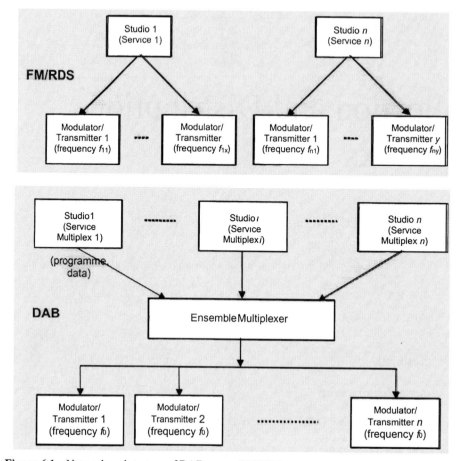

Figure 6.1   Network architecture of DAB versus FM/RDS

In general, co-operation among the different contributors to the DAB transmission signal has to be managed. This will normally be done based on contracts between service providers and the ensemble provider. Resources are allocated which comprise not only transmission capacities but also coding resources (unique identifiers) necessary for the signalling of multiplex configuration, service and other information to DAB receivers. When sub-divided into management categories usually applied to telecommunication networks, general management tasks to be realised in DAB networks are as shown in Table 6.1 below

Obviously the extent of technical support of these tasks depends on the specific DAB network scenario and its operational requirements. In a more static and regulated environment there is of course lower pressure to provide for technical support of management tasks than in the case when frequent changes are very common. Normally DAB service providers should be able to control a virtual ensemble of their own and to signal accompanying static and dynamic service information comparable to FM/RDS.

Table 6.1 General management tasks in DAB networks

Management Category	Task
Configuration	(Re-)Configuration of ensemble multiplex (Re-)Configuration of equipment (e.g. DAB mode, TII, physical interfaces, network and device delay)
Service	Co-ordination of service provider actions Monitoring and signalling of service information
Fault	Alarm and error signalling
Account	Admission of service providers Allocation of resources Billing
Security	Support of closed user groups Blocking of unauthorised control access

Concerning DAB networks, timing aspects have to be taken into account carefully. Not only do the single data streams of the service providers have to be synchronised, but also their actions have to be co-ordinated timely to prevent any recognisable impairments at the DAB receivers. Different network propagation delays between the ensemble multiplexer and the transmitter sites in the distribution network have to be compensated in order to exclude destructive RF signal superposition in the single frequency network. Therefore a common time and frequency reference is needed. Mostly the satellite-based Global Positioning System (GPS) is used for that purpose.

DAB standards dedicated to the collection network, that is the Service Transport Interface STI (EN 300797) as well as the distribution network, the Ensemble Transport Interface ETI (EN 300799), have been defined. These standards take into account the implementation of DAB networks based on public telecommunication networks. Specific error protection methods are defined to ensure safe and cost-efficient data transport from studios to transmitters. Although mainly data transport issues are dealt with, essential management tasks related to DAB operation are also specified by these standards. Consequently, the following sections are closely related to these standards.

6.1.2 Ensemble Multiplexer

The ensemble multiplexer forms the key element in the DAB signal generation chain. Its implemented functionality determines to a large extent how the collection and the distribution network can be operated, for instance which type of connections can be used and which DAB-related controlling and signalling possibilities are available for service providers. To prevent complete ensemble drop-outs, the requirements for error-free operation of the ensemble multiplexer are especially high.

Figure 6.2 presents a conceptual block diagram of the ensemble multiplexer. Incoming data streams have to be synchronised firstly and then usually demultiplexed because the preferably used STI data frames are designed to include several service components. According to the actual configuration, the remultiplex of service components has to be

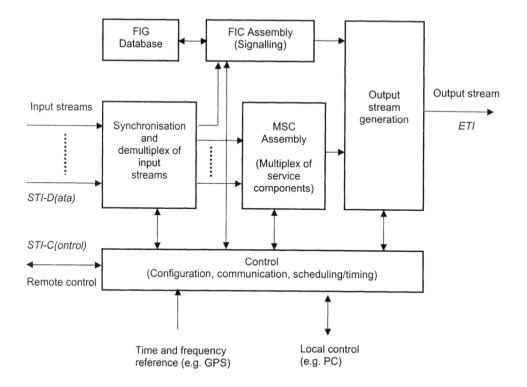

Figure 6.2   Conceptual block diagram of ensemble multiplexer

accomplished thereby creating the MSC part of the ETI output frame. With regard to the
24 ms frame timing, these processes take place synchronously, which means imposing a
fixed frame relation between the output and each of the contributing input data streams.

Concerning the signalling channel (FIC) which is part of the ETI output frame as well,
two different sources of data have to be used in general.

One is the local FIG database. It comprises not only the MCI FIGs corresponding to the
MSC configuration in use, but also FIGs defined by the ensemble provider (e.g. ensemble-
related information like frequency or region definitions). Additionally service information
FIGs delivered by service providers only once via control connection can be present there
(FIG files). It is necessary to feed all these FIGs periodically into the FIC. The respective
repetition rates are individually different and defined in (TR 101496).

The other source of FIGs to be transferred into the FIC are the STI input data streams which
can carry so-called FIG streams too, that is FIGs provided and, if necessary, repeated
by the service provider itself. These FIGs have to be transparently passed through the
ensemble multiplexer.

The whole FIC contribution to the 24 ms ETI frame has to be assembled into Fast
Information Blocks (FIBs) of 30 bytes length firstly. Concerning MCI, specific assembling
rules have to be met (TR 101496). The FIC assembling process is asynchronous by nature;
this means that in contrast to MSC contributions no fixed frame relation or delay exists

between FIGs delivered by service providers and their appearance in the ETI stream. Additionally consistency of FIG contents regarding resources allocated to service providers should be checked and ensured.

Because they are both reliable and cost efficient, the 2 Mbit/s (G.704) connections of public telecommunication networks are preferably used in the distribution network. It is the task of the ETI frame generation block to form transport data frames. Control information regarding final coding is included. ETI frames are normally error protected and also time-stamped in order to cope with delay compensation among connections of different length and type from ensemble multiplexer to the transmitter sites.

Beyond the necessary device configuration and subtle scheduling, the control block functions comprise a local control interface, for example a proprietary Application Programming Interface (API), as well as remote communication to service providers via STI. Management functions, as mentioned in the preceding section, are thereby supported including the data transfer from and to both remote sites and locally used tools supporting primary specifications of resources, configurations or FIGs and the monitoring of ensemble operation as well.

To ensure synchronised data transfer and control actions a common time and frequency reference is indispensable. GPS provides for both a reference clock and absolute time at all locations and is therefore preferably used. Taking into account the signal delay up to the transmission antenna, the ensemble time can be derived from the GPS time string. This time is signalled via the FIC to receivers and is also needed for timely co-ordination of activities in the collection network. Time-stamps are related to the 1 pps GPS signal.

It is obvious that this potentially large functional complexity of the ensemble multiplexer will lead to a wide variety of implementations. To ensure interoperability of devices from different manufacturers, compliance with the corresponding standards and implementation guidelines is most important.

6.1.3   Conceptual Broadcast Network Outline

Figure 6.3 presents the DAB conceptual broadcast network outline as described earlier. At the service provider side, in general a primary multiplex of service components (encoded audio, packet or stream data, fast information data channel) and service information will be provided and delivered via STI. Service providers in the collection network can also be replaced by sub-networks comprising further STI service providers.

Both the collection and the distribution network are dealt with in detail in sections 6.2 and 6.3. Concerning the delay in the distribution network, refer also to section 7.4. A complete implementation example is described in section 6.4.

6.1.4   Implementation Issues

Many aspects have to be considered in the course of planning and implementing a DAB broadcast system. With regard to collection and distribution networks some important aspects will be listed here briefly. The coverage area of a Single Frequency Network (SFN) is dealt with in Chapter 7.

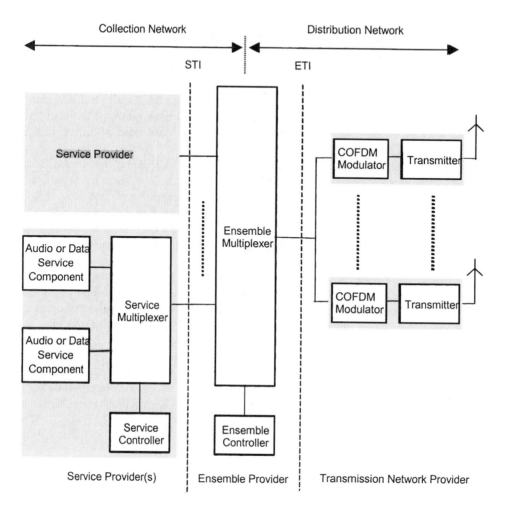

Figure 6.3   DAB conceptual broadcast network outline

From an operational point of view the following requirements are essential:

- Reliability of operation, based on technical stable networks and correct manual specifications created as much as possible "right by construction"
- Ease of use, for example tool support related to different user classes, for instance (pre-) specification in contrast to simple scheduling of configurations
- Automated operation, including scheduling of control actions, error and alarm reactions, remote operation
- Provision of flexibility, for example ability of service providers to control their part of ensemble resources based on constraints individually allocated by the ensemble provider who has to avoid erroneous interference due to the commonly used FIC
- Integration into existing infrastructure, for example reuse of available RDS data sources and connection to automated play-to-air systems (see Chapter 5)

- Cost efficiency, for example alternative use of telecom connections (for instance, use of bundled ISDN channels instead of 2 Mbit/s G.704 PDH lines by service providers).

Consequently, the following requirements for the respective equipment can be derived:

- Reliability, that is of course a high MTBF (maybe redundancy) but also supervision of regular operation including error and alarm handling
- Variety of (configurable) physical interfaces to support different types of connections
- Automated control, for example by means of scheduling functions integrated in service and ensemble controllers
- Technical support of management tasks, for example resource allocation
- Open system architecture supporting interworking in existing infrastructure (e.g. API)
- Standard compliance of equipment to ensure interoperability between devices of different manufacturers
- Ability to integrate non-standard devices (e.g. available non-STI audio codecs).

Of course, not all of these requirements will be of the same importance in any case. According to evolving requirements, evolution strategies are preferred. This comprises feedback and evaluation of practical experiences, too.

6.2   The Collection Network

6.2.1   The Service Transport Interface (STI)

The STI was originally defined within the Eureka 147 project and published by ETSI as (EN 300797). The standard is rather complex because it covers both a large variety of data transmission issues and also DAB specific management tasks related to the collection network.

The basic building blocks of an STI-based collection network are point-to-point connections established between devices supporting STI at their respective connectors. From a logical point of view the STI connection is sub-divided into a unidirectional data connection, mainly conveying the data streams to be broadcast, and a bidirectional control connection. Of course, this does not imply strict separation with respect to physical connections.

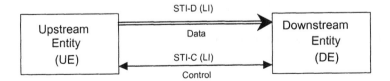

Figure 6.4   Logical STI connection

Regarding the single logical STI connection as shown in Figure 6.4, STI devices can be depicted in general as abstract entities which are distinguished according to the direction of the main data flow. The upstream entity delivers the data and therefore represents a service provider in every case. The entity at the receiving side of the data connection is called the downstream entity and can represent the ensemble provider or alternatively an intermediate service provider in the collection network.

The conceptual model of the STI is outlined in Figure 6.5. Beyond the separation of data and control, layers are introduced to reduce complexity and to provide for further internal interfaces.

At the top layer the basic data frame structure and control messages are defined. Consequently it consists of two logical interface parts: the data part STI-D(LI) and the control part STI-C(LI). Unlike broadcast data transport, error-free transport of control data is indispensable but not time critical.

Therefore an additional transport adaptation layer STI-C(TA) ensuring safe transport and of control data was defined. This includes the ability to address the respective entities, too. Based on the definition of a common generic transport frame, the physical interface layer STI(PI,X) standardises how STI should apply widely used physical interfaces. Thereby aspects of framing, forward error correction, delay compensation as well as electrical parameters and connectors had to be considered.

The different parts of the STI definition will be dealt with in the following sections.

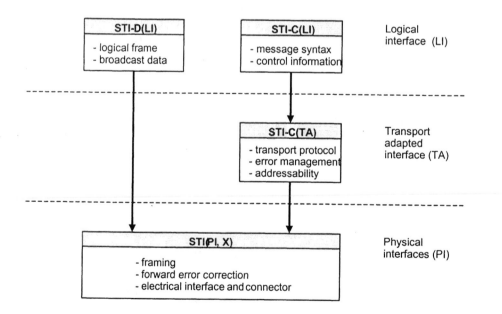

Figure 6.5   Layer structure of STI

6.2.1.1 The Logical Data Interface

The frame structure of the logical data interface STI-D(LI) is represented in Figure 6.6. To provide for flexible use, the STI-D(LI) frame is related to the common interleaved frame (CIF) as defined in (EN 300 401).

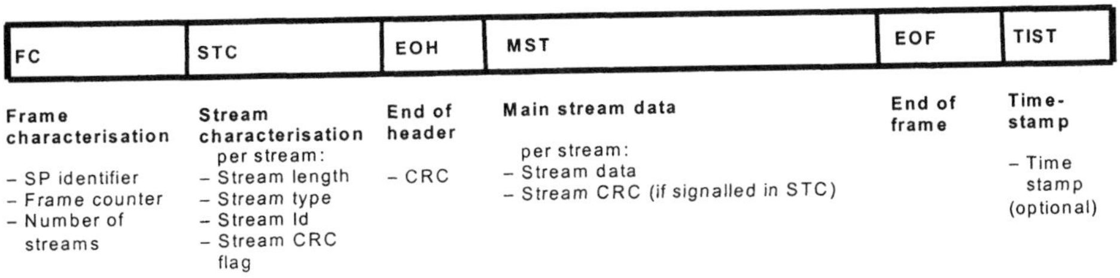

FC	STC	EOH	MST	EOF	TIST
Frame characterisation	Stream characterisation per stream:	End of header	Main stream data	End of frame	Time-stamp
– SP identifier – Frame counter – Number of streams	– Stream length – Stream type – Stream Id – Stream CRC flag	– CRC	per stream: – Stream data – Stream CRC (if signalled in STC)		– Time stamp (optional)

Figure 6.6   Frame structure of the STI-D(LI)

Apart from the source identification, the unique service provider identifier, each single frame can be identified additionally within time slots of 2 minutes by its modulo 5000 frame counter value. In general, the STI-D(LI) frame will comprise several streams each individually described in the stream characterisation field. Possible stream types are listed in Table 6.2.

Independent of the contents, two basic classes of streams have to be distinguished: continuous and discontinuous. Continuous streams have a fixed amount of data per STI-D(LI) frame. According to initial synchronisation they have to experience a constant delay being passed through the DE. Thereby synchronous transmission is ensured. All of the MSC-related stream types with the exception of the MSC packet mode contribution are of

Table 6.2   STI-D(LI) stream types

Stream	Type/Extension	Comment
MSC audio stream[*]	0/0	Independent of sample frequency
MSC data stream[*]	0/1	
MSC packet mode stream[*]	0/2	
MSC packet mode contribution	3/0	Needs to be adapted to sub-channel
FIC service information[*]	4/0	No MCI FIGs allowed
FIC data channel[*]	4/1	TMC, paging, EWS
FIC conditional access information	4/2	EMM and EMC FIGs (only an alternative to usage of MSC)
FIB (asynchronous insertion)	5/0	32 bytes per frame (FIGs and CRC)
FIB (synchronous insertion)	5/1	FIBGRID message defines insertion
In-house data	7/x	User definable (no broadcast)

[*] Basic stream types as defined in (TR 101xxx), see section 6.2.4.

that type. The latter requires pre-processing, that is an additional pre-multiplex step to merge several streams of this type and to form a complete MSC packet mode stream with a sub-channel bit rate of nx8 or nx32 kbit/s as required by (EN 300401).

Unlike continuous streams, discontinuous streams will not convey the same amount of data or data at all in each of the consecutively provided STI-D(LI) frames and, in general, data will not be delayed constantly passing through the DE.

The actual number of bytes per stream and frame is called stream length and is dynamically signalled in the stream characterisation field. Stream length equals zero if a stream is logically opened but has no data in the respective frame.

A stream identifier (stream Id) is used to refer to streams by STI-C(LI) configuration data. It should be noted that the stream data extraction process from STI frames received is performed under control of the current configuration at the ensemble multiplexer (or DE in general). Further stream type/extension coding values are reserved for future definition.

6.2.1.2   The Logical Control Interface

The logical control interface is based on ASCII-coded messages which can be transferred asynchronously. A bidirectional connection is assumed to support typical request response communication between the two entities.

The unified message format of STI-C(LI) consists of a seven-character command word followed by a three-character extension and an individual set of parameters. All fields are separated by blanks and the message is terminated by a semicolon. The whole set of messages and its usage are described in detail in (EN 300797). The standard also defines which kind of entity (upstream or downstream) is entitled to use a certain message and how to respond on its reception. For the sub-set definition of the message set with regard to STI implementation levels refer to section 6.2.4.

So-called data exchange sessions are used at logical interface level to provide for complete block transfer of logically related messages, for example an MCI configuration or a FIGFILE. In Table 6.3 a survey is given of the spectrum of STI-C(LI) messages. Each message class comprises a number of messages, specified for the different purposes.

Table 6.3   Message classes of STI-C(LI)

Message Class	Messages	Function	Message Example
ACTION	6	Control of reconfiguration	RCONFIG REQ name time framecount;
CONFIG	10	Ensemble configuration	CONFDEF DEF name <spec>;
FIGFILE	8	Static FIG data	FIGFILE REC <FIG data>;
FIBGRID	6	FIB insertion (into FIC)	FIBGRID REC <data>;
RESOURCE	27	Resource allocation	CHANCAP DEF MSC capacity;
INFORMATION	16	Information about status	COUNTER INF;
SUPERVISION	9	Error and alarm information	ALARMST DEF status time;

6.2.1.3   The Transport-adapted Control Interface

The transport adaptation is aimed at ensuring an error-free transmission of STI-C(LI) messages including the addressing capability. Unlike for audio channels repeated transmission is not a principal problem. Therefore STI-C(TA) consists of a TCP/IP-like transport protocol which can be replaced by real TCP/IP under certain conditions (see section 6.2.5). Differences from TCP/IP concern the reduced length of packet numbers and addresses.

Figure 6.7 illustrates how STI-C(TA) data are formed. This process is again based on a layered approach. The whole functionality could be assigned to the transport layer of the OSI reference model because the STI-C is an end-to-end connection between two STI entities. Nevertheless functionality analogous to lower OSI layers is used to describe the different functional parts.

At the receiving side the CRC of the data packet will be checked first. In case there is no error then the useful data will be submitted to the layer above. Otherwise the data will be ignored and resent by the source entity later on owing to missing acknowledgement. At the next layer the destination address of the packet is compared with the address of the receiver. If there is no match, the data packet will again be ignored. In the opposite case, the data are intended for the receiving entity and error-free transmitted.

Consequently the next layer will acknowledge its reception using the respective packet number under the condition that no packets have been lost in between. The acknowledgement requires sending a message back to source and can be combined with transmission of useful information. Finally the STI-C(LI) DATA, a sub-string of pre-defined length of the overall message sequence at source side, is passed to the logical layer.

It should be emphasised that the received data packet in general will not be aligned to one or more complete STI-C(LI) messages. Message detection and interpretation takes place in the resulting character string formed by appending the data packets consecutively as received. The overhead per transport packet is 32 bytes. The transmitting side behaves in the opposite way.

The protocol requires an explicit open/close of connections. Owing to the multiplex of logical connections corresponding to SAD/DAD addresses, a common STI-C channel can be used to communicate with several destinations.

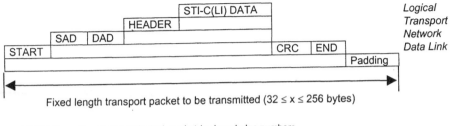

HEADER   Transport header incl. packet / acknowledge numbers
SAD/DAD   Start / Destination address
START/END   Delimiter characters

Figure 6.7   Layer structure and format of STI-C(TA) packets

This will be particularly useful for instance in the case of a common reverse channel from the ensemble multiplexer or in the case of multicasting the same forward channel from the service provider (e.g. together with STI-D via satellite).

6.2.1.4   The Physical STI Interface

Physically transported frames are all derived from a generic transport frame as shown in Figure 6.8.

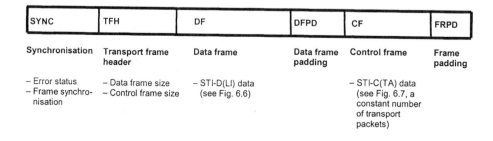

SYNC	TFH	DF	DFPD	CF	FRPD

Synchronisation	Transport frame header	Data frame	Data frame padding	Control frame	Frame padding

– Error status
– Frame synchro-
  nisation
	– Data frame size
	– Control frame size
		– STI-D(LI) data
		  (see Fig. 6.6)
				– STI-C(TA) data
				  (see Fig. 6.7, a
				  constant number
				  of transport
				  packets)	

Figure 6.8   Frame structure of the generic transport frame STI(PI,X)

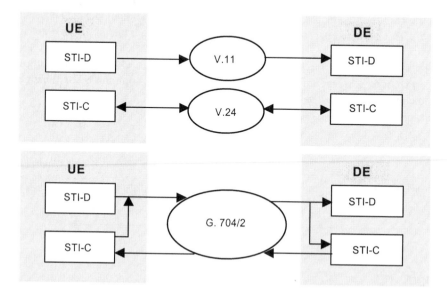

Figure 6.9   Physical connections of STI-D and STI-C (examples)

Compared to STI-D(LI) it contains the additional synchronisation field including an error status field, marking the severity of errors detected and set by the error-detecting unit. Further on, a transport frame header is inserted which signals the fixed size of the following data and control part. Either STI-D(LI) or STI-C(TA) or both can be transported depending on the respective field size described in the transport header.

The generic transport frame has to be adapted to the specific physical interface in use. Frame assembly details as well as connectors are described in (EN 300797). In total the standard defines eight physical interfaces. For synchronous transmission there are available the widely used 2 Mbit/s interfaces G.704 (with and without FEC as in ETI) and G.703.

Further interfaces concern bundled ISDN channels including delay compensation according to H.221 (J.52), V.11 (RS422), IEC 958 and WG1/2 from Eureka 147. V.24 (RS232) can be used for asynchronous, low-bit-rate transmission. Although it is appropriate in most cases to transport STI-D and STI-C within the same connection, the standard provides possibilities of transmission using separate connections. Figure 6.9 exemplifies both cases.

6.2.2   Network Architecture

According to (EN 300797) the collection network is logically based on point-to-point connections forming a tree-like hierarchy of entities. Figure 6.10 outlines this by means of an example. It should be recalled that diverse physical connections between entities can be implemented.

Each entity can be classified by a type according to its role in the network (TS 101860). As already introduced in section 6.1, the essential roles are service or ensemble provider.

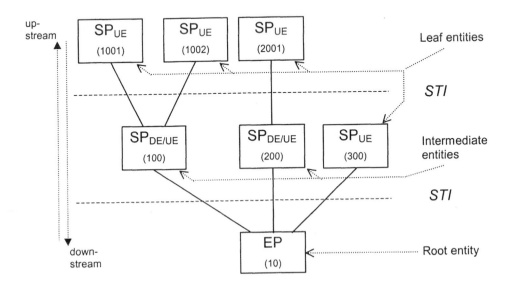

Figure 6.10   Tree-structured DAB collection network (example)

Based on implications to specific STI functionality to be implemented, service providers are further distinguished into those who are in the UE position and those who are situated intermediately in the tree, that is those who are both UE and DE (see also section 6.2.4).

As outlined in Figure 6.10 each entity needs to be identified uniquely by a service or ensemble provider identifier. These identifiers are used for the addressing of control messages and marking of logical data frame sources as well. No specific numbering scheme is pre-defined. The total of transmission and coding resources allocated to an entity have to be shared between this entity and all of the entities connected to its inputs. For instance, in Figure 6.10, resources allocated to SP 100 comprise resources reserved for local use by SP 100 itself (e.g. due to service components or information inserted by non-STI sources at this level) and those propagated to its subordinated SPs, that is to SP 1001 and 1002.

At least during the migration phase real networks may additionally comprise non-STI SPs (see also section 6.2.4). Section 6.4 presents a corresponding example.

6.2.3   Network Operation

The requirement of both flexible and safe regular operation of the complete DAB broadcast network constitutes a challenge due to the complexity of the distributed real-time system. In general, service providers should be able to operate independently of each other but co-ordinated by the ensemble provider, based on respective contracts.

Assuming collection network has been established according to section 6.2.2, mainly STI-C(LI)-based management tasks and subsequent actions have to be considered concerning dynamical operation. STI-D stream processing is consequently controlled. For simplicity, the focus will be on the most important first hierarchical level of the collection network, that is EP and directly connected SPs. The STI defines an interface of machine-readable data and no user interface for operators. Consequently, in practice there will be a need for additional tools performing the translation tasks from manual specifications to STI control messages or FIGs and vice versa.

In order to enable service providers to start broadcasting, the ensemble provider has to provide some prerequisites. The control connections to the service providers have to be opened using the provider identifiers as negotiated. Via these connections information about individually allocated resources could be transmitted to the service providers if implemented. Otherwise this information has to be passed to them outside of STI. Essential data, which can be delivered via STI-C by RESOURCE messages, comprise:

- Identifiers of sub-channels, services, FIC data channels and linkage sets
- Transmission capacities of MSC and FIC
- Maximum number and bit rates of specific stream types
- Restrictions concerning usable FIGs (regarding FIG type/extension)
- Entitlement and parameters to send announcements.

Based on there, service providers are able to specify and check activities of their own.

Further on, an empty ensemble can be generated, which means no service components are included before the first reconfiguration initiated by a service provider. Nevertheless basic ensemble-related FIGs can be already sent as defined by the ensemble provider. Essentially this FIG set will consist of

- Ensemble identification and label (FIG 0/0, FIG 1/0)
- Date and time (FIG 0/10)
- Ensemble-related tables to be used, country with local time offset (time zone) (FIG 0/9)
- Frequency lists (FIG 0/21)
- TII lists (FIG 0/22)
- Regional definitions (FIG 0/11).

Service providers will normally start transmission by performing a reconfiguration of the virtual ensemble allocated to them, for instance by means of RESOURCE messages as mentioned above. Figure 6.11 shows this process in chronological sequence.

First of all the multiplex configuration has to be defined in the frame of a data exchange session. Therefore the first message block will be enclosed by CONFDEF DEF and CONFDEF END messages. The CONFDEF DEF message is the header of the block providing the name of the configuration and numbers of messages of specific type included.

The body of the message block contains the following definitions (one message per item):

- Sub-channels to be used for transmission of service components including identifiers and protection levels (SUBCHAN DEF)
- Service components (CMPNENT DEF) including service component types and corresponding stream and sub-channel identifiers (for both MSC and FIDC), also including packet addresses in the case of packet mode service components
- Service definitions (SERVICE DEF) comprising one or more service components and based on the specifications given before
- Further optional definitions, for instance to specify by name that predefined FIG files (signalling data) have to be enabled at reconfiguration instants (USEFIGF DEF) or that FIG streams out of STI-D have to be inserted into FIC (USESTRM DEF).

Firstly, the ensemble provider needs these data for generation of MCI FIGs to signal the multiplex on-air. Secondly, these data are necessary for controlling stream extraction from STI-D to form the MSC and to enrich the FIC in case FIG streams are in use.

The messages received will be checked for plausibility and compliance to the allocated resources. If the checks are passed, the configuration is stored and can be referred to from there on. Otherwise an error message is sent back to the service provider. How many configurations an ensemble provider is able to store depends on implementation and usage. Service providers can be informed about that via STI-C. Next the service provider will request a reconfiguration from the ensemble provider specifying the name of the configuration and the execution time wanted (RCONFIG REQ). To ensure seamless reconfiguration the time specification ($<t5>$ in Figure 6.11) must be frame related. It is therefore sub-divided into a UTC value uniquely addressing 2 minutes out of 24 hours and a data frame count value specifying the respective 24 ms frame within the 2 minutes.

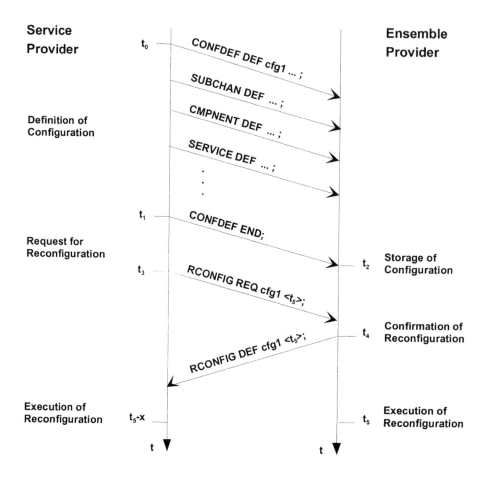

Figure 6.11   Control message flow for definition and activation of virtual ensemble configuration
(example)

According to (EN 300 401) two subsequent reconfigurations must have a time distance of at least 6 seconds. Because there is no co-ordination required between service providers to agree on reconfiguration times, only the ensemble provider is entitled to define the instant of execution. Thereby the latter is able to merge the requirements of several providers in such a way that only execution times equal to or later than the requested ones are allocated.

Correspondingly the ensemble provider will check that a correct configuration of the received name has been defined previously by the service provider and will define the execution time for reconfiguration as close as possible to that requested. This information is sent back to the service provider (RCONFIG DEF). From now on a reconfiguration is pending at the ensemble multiplexer which will lead to the corresponding on-air signalling of the next ensemble configuration starting automatically about 6 s before reconfiguration. Per provider, only one reconfiguration can be pending at a time.

Assuming that the service provider does not cancel the reconfiguration request right in time (RCONFIG CAN), it is expected that contributions are delivered via STI-D precisely timed and in accordance with the activated configuration. Otherwise error messages would be generated by the ensemble provider. Some further remarks are needed regarding the right instance to change streams or bit rates. As depicted in Figure 6.11 the service provider needs to know the propagation delay x of the STI-D frames, that is the time offset between departure of a frame at the service multiplexer and the appearance of its contents in the respective ETI frame leaving the ensemble multiplexer. By means of the message COUNTER INF a service provider can ask the ensemble provider about the frame-counter-related time difference and has to take this into account to support seamless reconfiguration. Moreover, decreasing the bit rate of a sub-channel needs to be accomplished 15 frames in advance to comply with the 16 frame interleaving process of the on-air signal.

Provision of service and ensemble information is essential. As mentioned earlier, there are two basic mechanisms to be used in conjunction or alternatively: FIG streams (transported in STI-D) and FIG files (transported in STI-C). The main difference concerns the responsibility for FIG repetition. Required repetition rates are defined in (TR 101496). In the case of FIG streams, which are transparently passed into the on-air FIC, this must be performed by the source, for example by service provider. Otherwise the ensemble provider is responsible for repetition of FIGs.

It should be emphasised that special treatment of so-called list FIGs is necessary. Mainly information is concerned with

- Service linking sets
- Region definitions
- Frequency lists
- Transmitter identification lists
- Announcement support
- Other ensemble services
- Satellite database.

These FIGs comprise small databases. Well-defined parts of them should be updated selectively in order to avoid time-consuming recovery of complete data by receivers in case of changes. That is, with regard to key data fields the change event concerning a data subset has to be signalled by means of specially defined FIGs sent for 6 seconds between the old and the new data versions (SIV/CEI, see also TR 101496).

Additionally or alternatively FIG files can be used to signal supporting information. As with ensemble configurations, the corresponding file, which consists of one or more FIGs, has to be transferred to the ensemble multiplexer first. It would be checked and stored analogously under its name.

Further on, two alternative activation mechanisms are standardised for FIG files. As mentioned above, the FIG file to be activated at the instant of reconfiguration should be named by means of a USEFIGF DEF control message which is part of the respective ensemble configuration. Alternatively, FIG files can be activated or deactivated explicitly using FIGFILE SEL(elect) or DES(elect) messages respectively at arbitrary time points. Thereby activation/deactivation can be triggered in principle either based on pre-defined schedules or derived from events like keystroking by a moderator. Repetition of FIGs as part of FIG files including the SIV/CEI treatment is up to the ensemble provider.

Intermediate service providers exist in hierarchical collection networks (Figure 6.10) consisting of more than one level of service providers. As a downstream entity they should communicate with its upstream service providers like an ensemble provider. In the opposite direction, as an upstream entity, it should behave like other service providers (leaf entities). Consequently, intermediate service providers have to share resources with and act on behalf on their subordinated service providers.

Because STI covers only management issues very close to DAB, there is a need for support beyond STI to operate real networks. For example, section 6.4 will briefly cover such requirements and their practical implementation.

Important requirements are:

- Automation of operation (scheduling)
- Coding support for ensemble configurations and FIGs guided by allocated resources
- Event logging
- Alarm handling
- Connection to existing infrastructure (e.g. studio operation system, databases/sources).

The problem of interoperability due to devices from different manufacturers who may each implement different subsets of the STI is considered in section 6.2.4.

6.2.4   STI Implementation Levels

As demonstrated above, the STI standard (EN 300797) is rather complex and therefore the question arises how the interoperability of equipment produced by different manufacturers can be ensured. It is quite reasonable that not all STI-compliant devices have to provide for the complete functionality as defined in the standard. Especially owing to the different extents of dynamic operation of the specific collection network and the number and kind of providers involved there will of course be varying requirements leading to different implementation costs too.

Faced with this situation, a WorldDAB task force derived subsets of functionality from the standard. It was assumed that the standard itself should not be changed. A technical report defining "STI Levels" has been standardised by ETSI (TS 101860).

Analysis has shown that definitions of basic stream types to be processed as well as management functionality related to the STI control functions are most important considering useful sub-sets of functionality. Therefore sub-division of functionality has been done with regard to these critera. Three hierarchical levels of STI functionality have been defined. This means that a higher STI level fully encloses the lower ones. But even the highest level, the third level, does not comprise the whole functionality of (EN 300797). Some special functions are declared to be optional and level independent.

Also physical interfaces are beyond the scope of the STI level definition because it is much easier to reach interoperability between devices on this level than on a functional level. A wide variety of terminal adapters and other interface converters can be used for that purpose. Table 6.4 outlines the STI levels as defined in (TS 101860).

Table 6.4   STI implementation levels

Level	Interface	STI-D(LI) Stream Types	STI-C(LI) Messages	Comment
1	Restricted		(no control channel)	Local control proxy for UE at DE needed
2	Regular	Processing of basic stream types: MSC audio MSC stream data MSC packet data FIG (SI) FIG (FIDC)	ACTION, CONFIG, FIGFILE, INFORMATION, SUPERVISION	Seamless dynamic re-configuration initiated by remote SP (UE), FIG file-based signalling, status and error messages
3	Advanced		RESOURCE	Resource allocation and consistency checks, more status and error messages
	Options	Processing of other stream types: FIB, PMC, FIG(CA)	FIBGRID Reconfiguration enforcement of UE by DE	Selection of physical interfaces according to STI standard by both users and implementers

Implementation of the control channel is not required at STI Level 1. Only STI-D has to be supported regarding basic stream types. In contrast to the ensemble provider, or a downstream entity in general, service providers are only enforced to generate at least one out of the basic stream types. Implementations at the downstream side will be similar to those aimed at supporting non-STI providers (refer to section 6.2.5).

To be compliant with STI Level 2 the availability of the control channel is assumed and processing of a basic set of STI-C(LI) messages has to be implemented. This leads to the ability for seamless dynamic reconfiguration initiated by the remotely situated service provider. FIG files can be used for FIC signalling. Both FIG file activation mechanisms as described in section 6.2.3 have to be implemented.

The advanced STI Level 3 additionally provides for ensemble co-ordination by means of RESOURCE messages. Nearly the full functionality of STI-C(LI) has to be realised to be compliant. Upstream entities working at the same level get useful information via STI-C to guide their specification process and to make it right first time. At downstream entities consistency checks will be carried out on all directly connected upstream entities based on the respective resources allocated. This concerns supervision of used stream rates and transmission capacities as well as the checking of FIG contents. Severe interference between service providers can thereby be excluded.

For example, this relates to service and service component identifiers used in a couple of FIGs to signal service labels, programme type, announcement support and so on. Also linkage and announcement provision data (e.g. cluster Id, announcement type and, if applicable, sub-channel Id) should be checked. In case of errors or inconsistencies the generation of STI-C(LI) error messages takes place, for instance STERROR DEF messages pointing out respective stream Id and error type.

Optional functions which are beyond the scope of the STI level definition concern mainly

- Ensemble output of time, respectively frame, window related FIGs as needed for low-power-consuming receivers dealing with paging or emergency warning systems or for entitlement messages to be signalled in FIC to support conditional access
- MSC sub-channel contributions, as PMC, which need to be pre-processed before mapping into ETI frames
- Use of FIB instead of FIG streams leading to specific FIC assembling requirements
- Enforcement of upstream entity reconfigurations by the downstream entity.

These functions, if needed, have to be negotiated between users and implementers individually because they cannot be assumed as supported by STI devices in general.

STI levels should be assigned to single interfaces (input and/or output) of STI devices. Connecting interfaces of different levels by an STI connection is possible. In order not to lose functionality, at the paths from the leaf entities to the root entity the following rule should apply to each point-to-point STI connection: the STI level of the upstream entity should be lower than or equal to the STI level of the downstream entity. In case this does not hold and the upstream entity uses an STI-C(LI) message not implemented at the downstream side, the downstream entity should answer with a message informing about the reception of an unknown message.

6.2.5   Integration of Non-STI Service Providers

DAB collection networks according to Figure 6.10 are exclusively based on STI and therefore do not take into account service providers not compliant with STI. Integration of these so-called non-STI service providers supports migration to DAB and is motivated by the following requirements.

At least in the introductory phase of DAB there are broadcasters who intend to start DAB transmission without substitution of available but not STI-compliant equipment. They will accept some operational restrictions instead. Secondly, in special cases of co-operation among ensemble providers the ability to extract service components from one ensemble output data stream (ETI, see section 6.3) and insert them into another ensemble could be useful.

A principal solution can be based on splitting up both the control and data parts included in the STI. The non-STI service provider owns no STI control channel and delivers only the service component data, in the simplest case only one DAB-formatted audio stream according to (EN 300401), using arbitrary physical interfaces supported at the input of the receiving downstream entity. No FIG streams can be delivered. Necessary configuration and service signalling data have to be provided by the downstream entity using its local control interface. In other words, the downstream entity completely takes over ensemble control including the signalling of service information as proxy for a non-STI service provider. This is close to support of service providers at STI Level 1.

Surely some restrictions will be imposed regarding regular operation where service and ensemble provider typically belong to different organisations and where normally an ensemble provider will not be able to perform frequent and timely co-ordinated changes in service signalling or dynamic reconfigurations on behalf of a remotely situated non-STI service provider. Therefore, application of this approach will be restricted to special cases of static kinds of operation without dynamically signalling service information.

The network example given in section 6.4 shows that the approach described above has been implemented successfully.

6.2.6   *Advanced Features and Further Development*

In the early stage of DAB introduction, not all features defined in the respective standards are of the same importance and will be implemented immediately. The most important features related to collection networks have been described earlier, also taking into account different STI implementation levels.

Based on the changes in the broadcast landscape in general, and experiences gathered on collection networks, implementations are developed with respect to three categories:

- Use of advanced features already standardised
- Evolution of standards, especially the STI standard
- Evolution outside DAB standardisation.

The first category mainly comprises features such as

- Extended use of satellites in the collection networks (multicast of STI data)
- Dynamic change of local services and respective signalling of local service areas
- Dynamic post-processing of received signalling data by the ensemble provider, for example to form ensemble preview on programme types
- Use of the FIC overflow channel (Auxiliary Information Channel, AIC)
- Provision for time window related FIC output by (paging, conditional access, EWS).

Evolution of the STI standard could mean for instance:

- Acknowledgement of every transaction executed to support closer supervision
- Additional information messages, for example to ask for FIG files which are currently active
- Provision of ensemble-related global signalling data as, for instance, table, country and regional definitions to upstream entities thereby supporting specification of related FIGs

- Support for cross-signalling between co-operating ensemble providers or even cascading of ensembles
- Standardisation of conformance tests for STI devices.

All in all, open and reliable system solutions providing for easy and seamless adaptation to the embodying infrastructure as well as offering trade-offs between functionality and expenditure will be developed stepwise.

6.3   The Distribution Network

6.3.1   The Ensemble Transport Interface (ETI)

Originally defined within the Eureka 147 project, the ETI was published by ETSI (EN 300799). It is intended to be used for the distribution of ensemble data from the ensemble multiplexer to the transmitting sites of the single frequency network (SFN).

The ETI is conceptually based on layers differentiating between the logical interface (LI) and, at the physical layer, network-independent interfaces (NI) and network-adapted interfaces (NA). For physical transport in public telecommunication networks, it has been decided not to exceed the bit rate of 2 Mbit/s with regard to cost efficiency. Consequently, the process of convolutional encoding, which leads to higher ensemble data rates of 2448 kbit/s (DAB transmission mode III) or 2432 kbit/s (DAB mode I, II, IV), must be shifted from the ensemble multiplexer to transmission sites. The data to be broadcast are transmitted uncoded via ETI and normally another means of error protection is needed. Forward Error Correction (FEC) based on Reed Solomon (RS) block coding has therefore been defined in ETI(NA). In the following, the different ETI layers will be described in more detail.

The frame structure of the logical data interface ETI(LI) is presented in Figure 6.12 and corresponds to those of STI-D(LI) as presented in Figure 6.6. As STI-D(LI), the ETI(LI) frame carries data that are related timely to the 24 ms Common Interleaved Frame (CIF) formed in the channel encoder at the transmission site.

The frame characterisation field comprises the lower part of the complete frame count (modulo 250, i.e. at 6 s periodicity) and a flag to point out that FIC data are present at the beginning of the main stream data part. The amount of FIC data, if present, is determined by the DAB mode signalled as well (three or four FIBs with 32 bytes each depending on the transmission mode). Further on, the number of streams contained, and the overall frame length, is described in the frame characterisation field. The frame phase (FP) consists of a modulo 8 counter, incremented at each frame, and controls the TII insertion in the channel encoder. On starting the ensemble multiplexer it must be ensured that FP zero is aligned to CIF count zero.

Each stream to be broadcast in the MSC is specified by its single stream characterisation (SSTC) which thereby commands the channel encoder, see section 6.3.3. As depicted in Figure 6.12, the start address in MSC (in CU, 0 ... 863), the sub-channel identifier as well as service component type and protection level to be applied by the channel encoder are given in addition of the length of the respective stream data in the MST field. Of course, data included in the SSTC have to be compliant with those signalled via FIC (i.e. the MCI).

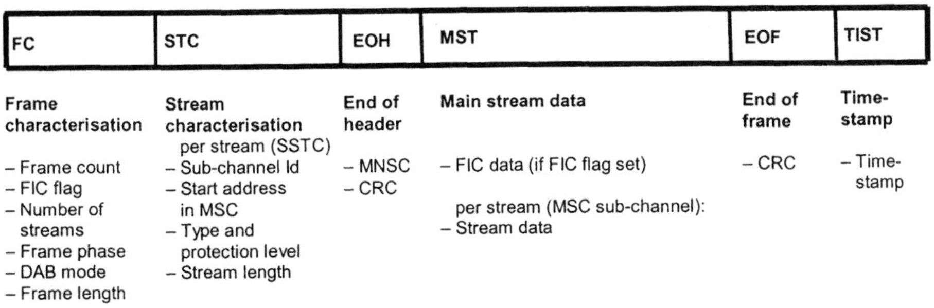

FC	STC	EOH	MST	EOF	TIST
Frame characterisation	Stream characterisation per stream (SSTC)	End of header	Main stream data	End of frame	Time-stamp
– Frame count – FIC flag – Number of streams – Frame phase – DAB mode – Frame length	– Sub-channel Id – Start address in MSC – Type and protection level – Stream length	– MNSC – CRC	– FIC data (if FIC flag set) per stream (MSC sub-channel): – Stream data	– CRC	– Time-stamp

Figure 6.12   Frame structure of Ensemble Transport Interface ETI(LI)

In contrast to STI-D(LI), only streams constituting complete MSC sub-channels can be transported using ETI(LI).

The subsequent end of header (EOH) field comprises not only a CRC for error detection but also 2 bytes of the multiplex network signalling channel (MNSC) to be used for management purposes (see section 6.3.3).

Frame data are completed by a further check sum of the MST contents and the ETI(LI) time-stamp. This time-stamp is thought for managing delay compensation in the transport network and defines the notional delivery time of the frame at the input of the channel encoder. Time is represented as (always positive) offset to a common time reference, mostly the one pulse per second (1 pps) of GPS. Time resolution is 61 ns. But, in most cases, accuracy will be determined by the accuracy of the time reference. An additional time-stamp is defined at the NA layer (see below).

At the physical layer, ETI(NI) is defined without error protection and therefore is only applicable for restricted use, such as for local connections or test purposes. To form an ETI(NI) frame, frame sync pattern and error status (analogous to STI generic transport frame, see Figure 6.8) as well as frame padding bytes (according to the bit rate used) are added to the ETI(LI). The interfaces ETI(NI,V.11) and ETI(NI,G.703) are defined in (EN 300799) to support RS432 and PDH-based connections respectively.

For operational distribution networks, ETI(NA) is normally used owing to the provision of FEC. The interfaces defined in (EN 300799) mainly concern two versions of interfaces compliant to G.704, but adaptation to PDH based on the first hierarchical level of 1544 kbit/s instead of 2048 kbit/s is given as well.

With respect to G.704 the mapping of 24 ms ETI(LI) frames into the corresponding multiframes of PDH is defined. There is a trade-off between redundancy used for FEC and the maximum useful bit rate available.

The first version, ETI(NA,G.704)$_{5592}$, provides for 5592 bytes of useful data (ETI(LI) data) corresponding to 1864 kbit/s. Further capacity up to the maximum bit rate of 1920 kbit/s is occupied by 40 kbit/s allocated to FEC and 16 kbit/s used for management and supervision purposes. These data include block counts used for frame synchronisation, a further time-stamp related to an NA link and defining the frame delivery time at the output of the respective converter from the NA layer to the NI or LI layer as well as a Network-adapted Signalling Channel (NASC) (see also section 6.3.3).

The respective figures of the alternative, the ETI(NA,G.704) $_{5376}$, are 1792 kbit/s for LI data, 112 kbit/s for FEC and again 16 kbit/s for management.

6.3.2   Network Architecture

According to (EN 300799), the DAB distribution network is unidirectional and logically based on point-to-multipoint connections. Figure 6.13 outlines this principle.

Because the radiated signals in the SFN have to be synchronised not only in frequency but also in time with tolerances of a few microseconds, automatic delay compensation is one of the most important issues. Therefore GPS receivers are usually located at each site involved to cater for the common time reference. Using the ETI(LI) time-stamps, it can be ensured that the overall delay throughout the transport network is always the same for every path between the ensemble multiplexer and a transmission site. Together with the information about the constant transmitter delay the ensemble multiplexer is able to evaluate the right time to schedule service providers and to signal time in FIC with regard to absolute time outside of the DAB network. See Chapter 7 (section 7.4.3) for details about delay definitions and treatment.

Network adapters as depicted in Figure 6.13 are needed to convert ETI(LI) into ETI(NA) and vice versa. This functionality is often integrated in the respective devices. The additional NA time-stamp allows in principle for the retiming of NA links, for example in the case of cascading ETI devices. Although support for distributed multiplexing would be useful in some special scenarios, this leads to considerably extra effort regarding synchronised actions and signalling among the sites involved and is therefore not considered further here.

The transport network in between the ETI network adapters can of course make use of further bearer services than PDH as long as G.704 payload can be transported transparently. For instance, additional terminal adapters to comply with SDH or ATM allow for the use of the corresponding networks. Satellite transmission or radio links are used alternatively to terrestrial connections too.

6.3.3   Network Operation

In the basic configuration, operation of the distribution network has to be performed fully automatically under control of the ensemble multiplexer. In contrast to the collection network, control is straightforward without response from transmission sites via ETI due to unidirectional communication. The basic configuration of the distribution network can be taken from the ensemble multiplexer using the signalling channels (MNSC, NASC, see also section 6.3.1) to channel encoders.

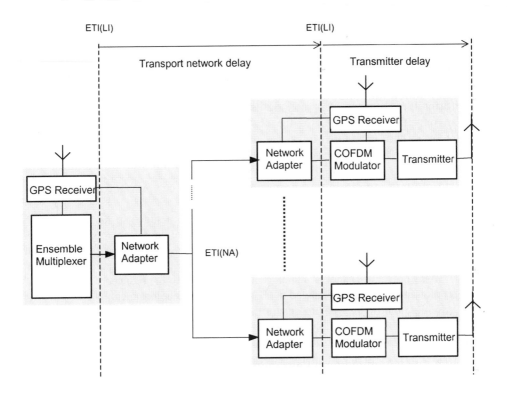

Figure 6.13 Typical DAB distribution network

The MNSC provides for 2 bytes per 24 ms frame, that is 667 bit/s at maximum. Messages for frame synchronous and asynchronous signalling are defined formally in Annex A of (EN 300799). Frame synchronous signalling can be used to transfer time information. Important applications of asynchronous signalling concern pre-definition of the individual TII and transmitter offset delay (see also section 7.4.3).

In the case when ETI(NA) is applied, the NASC can be used additionally to communicate between ETI(NA) network adapters. It has a capacity of 24 bytes per 24 ms, that is 8 kbit/s. As for the MNSC, frame synchronous as well as asynchronous signalling have been defined in general, but no application is pre-defined so far.

During regular operation dynamic delay compensation and especially reconfiguration of the MSC controlled via ETI(LI) have to be managed automatically. Delay compensation requires careful definition of delay figures and time reference in the network (see above) and will then work reliably.

Dynamic reconfiguration needs control of both the channel encoder and the receiver operated in the coverage area of the respective SFN. Normally, changes have to be accomplished inaudible to the listener and therefore require careful co-ordination.

Figure 6.15 below outlines how this can be performed based on frame-related actions carried out by the ensemble multiplexer, see also (EN 300401) and (EN 300799).

The signalling of multiplex reconfiguration starts in advance in the so-called preparation phase lasting up to 6 seconds. During this time the ensemble identification (FIG 0/0) is extended by a byte signalling the instant of planned reconfiguration in terms of the frame count (occurrence change). Additionally the type of reconfiguration (sub-channel or service organisation or both will be changed) is signalled by this FIG at each fourth frame.

In parallel, the current and the next MCI will be signalled at least three times to the receiver using the respective FIGs marking its scope by means of the current/next flag. In particular, only the relevant part of the next MCI, which depends on the type of re-configuration in turn, is required to be signalled. For instance, only the next sub-channel organisation FIGs (0/1) has to be sent in case the reconfiguration is restricted to that part of the MCI.

Service information related to the next configuration can be optionally provided also in advance during the preparation phase.
The channel encoder is controlled by means of the SSTC included in the header of the ETI(LI) frame. As depicted in Figure 6.14, it is important how a certain sub-channel changes in the course of reconfiguration. Owing to the time interleaving process, stream data delivered via ETI(LI) will be spread over 16 consecutive CIFs formed by the channel encoder. Therefore, sub-channels which will be removed or reduced in capacity at reconfiguration instant must be changed in ETI(LI) 15 frames before then.

  The ensemble multiplexer has to ensure this by of communicating with the respective service provider(s) and taking into account resource allocation. Bit error measurements aimed at single sub-channels, full transmission channel or telecommunication lines are usually based on pseudo-random binary sequences (EN 300799).

6.4   Example of Implementation

As shown in previous sections, well-done design of the entire network for the conveyance of all necessary contributions is a prerequisite for successful DAB operation.

In the phase before the introduction of regular operation, it was necessary to prove the operability of a real broadcast network based on the latest DAB standards. Therefore Deutsche Telekom established a partnership with Fa. Audio Video Technologies, Fraunhofer Institute Integrated Circuits, and Fa. Rohde & Schwarz with the aim of achieving and testing a complete workable solution.

In 1999 a field test in Berlin proved that the designed system and its components were working properly. On the basis of that field test a simple example of implementation focusing on the more challenging collection network will be presented in the following sections, see also (Nowottne, 1998) and (Peters, 1999).

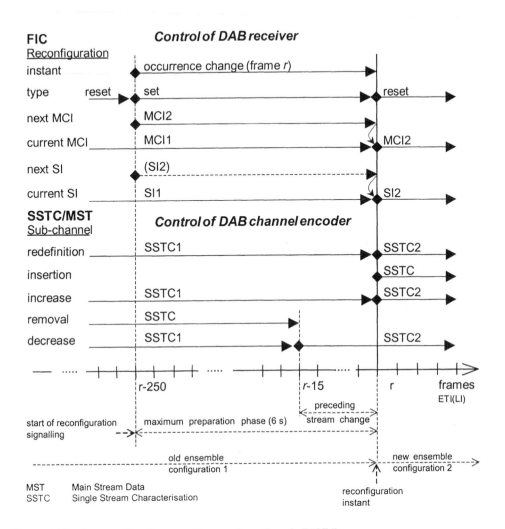

Figure 6.14   Commanding the ensemble reconfiguration via ETI(LI)

6.4.1   Operational Scenario

In the trial, STI service providers had to be allowed to operate their own service contributions and supporting signalling. The extent of the latter was comparable to FM/RDS, that is both static information such as service labels, programme types and announcement support, but also dynamic changing information such as for instance supporting traffic announcements via STI. Decentralised initiation of reconfiguration was needed to start broadcasting. To prevent failures, all specifications from the STI service providers had to be based on STI resource allocations provided via STI by the ensemble multiplexer and had to be supported by appropriate tools. Additionally, non-STI service providers had to be integrated into the collection network.

Figure 6.15 provides a diagram of the STI-based collection network as used in the field test. For instance, two STI service providers operate audio encoders MAGIC/STI for the generation of audio streams, optionally including insertion of PAD. Other service providers feed their service components to the ensemble multiplexer without using STI. Moreover, externally provided packet mode data stream, delivered in STI-D format, can be fed to the ensemble multiplexer directly or inserted by the MAGIC/STI service multiplexer. The MAGIC/STI interface to the ensemble multiplexer corresponds to STI(PI,G.704/2) and comprises both STI-D and STI-C with transport adaptation. As the ensemble multiplexer the DM001/STI by Rohde & Schwarz was used. MAGIC/STI as well as DM001/STI are connected via RS232 with PCs, working as service controller and ensemble controller, respectively.

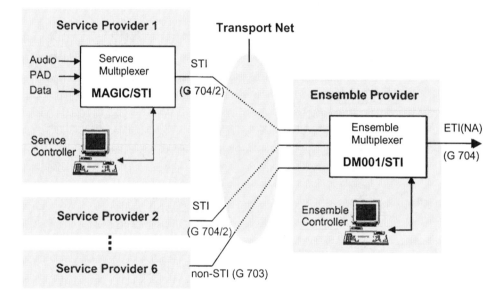

Figure 6.15   Structure of the collection network

As required, the system is also able to cater for non-STI service providers. In this case the ensemble provider takes the role of a proxy relating configuration and signalling.

6.4.2   The Service Provider Profile

The efficient treatment of resource allocation and consistency checking in an operational collection network is based on so-called Service Provider Profiles (SPPs). This has to be done by the ensemble provider in arrangement with service providers.

Each service provider is allowed to manage services independently within the constraints of the SPP. It is the task of the ensemble multiplexer to supervise the service provider's actions to ensure impact-free running. The management processes are performed and monitored by means of the STI control part, on which the service controller and ensemble multiplexer communicate.

Figure 6.16a outlines the contents of the applied SPP, as stored within the database of the ensemble multiplexer. These data comprise information on the service provider's address and basic entitlements, about capacity and coding resources, as well as other parameters.

Figure 6.16b gives an idea of the way it works. It has to distinguish between central specification and decentralised application. SPPs specified and stored at the ensemble provider are downloaded to the service providers and define their frame of action. The ensemble multiplexer supervises their operation, responds to change requests and carries out the changes if possible. Otherwise it rejects requests or reacts with error messages.

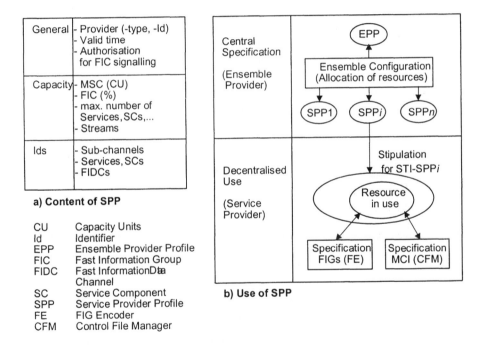

a) Content of SPP

CU   Capacity Units
Id   Identifier
EPP   Ensemble Provider Profile
FIC   Fast Information Group
FIDC   Fast InformationDta
          Channel
SC   Service Component
SPP   Service Provider Profile
FE   FIG Encoder
CFM   Control File Manager

b) Use of SPP

Figure 6.16   Management by means of service provider profiles

6.4.3   Equipment in Use

Two basic devices form the hardware basis of the introduced collection and distribution network: the MAGIC/STI Audio Encoder and Service Multiplexer by Fa. Audio Video Technologies and the DAB Ensemble Multiplexer DM001/STI by Fa. Rohde & Schwarz complemented by their Frame Decoder FD1000 for monitoring and analysis of both STI and ETI.

The structure and components of the Ensemble Multiplexer DM001/STI are depicted in Figure 6.17. The common clock and time reference is provided by an external GPS receiver. The DM001/STI is connected via an RS232 interface to its ensemble controller. Using this computer-based device the ensemble controller software, running under WindowsNT, supports STI-C-like communication with the multiplexer.

It is possible for the operator to define and download complete ensemble configurations each consisting of an EPP (Ensemble Provider Profile), several SPPs and all other ensemble-relevant parameters. To illustrate operation, the steps necessary to start broadcasting from the very beginning at the ensemble multiplexer are listed in the following:

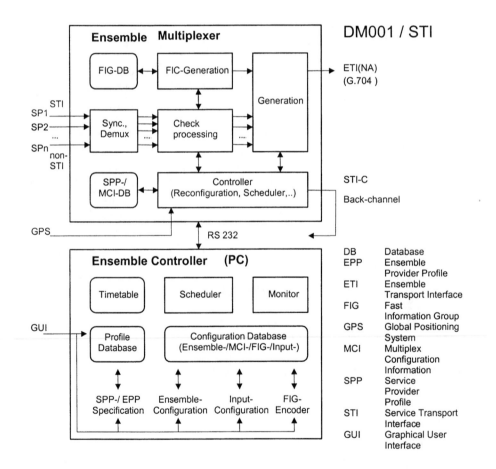

Figure 6.17   STI solution at the ensemble provider

1. Basic configuration of the DAB ensemble multiplexer, including
   - definition of DAB mode and external clock source (2.048 MHz from GPS)
   - definition of transport network and transmitter delay
   - definition of ETI type, for example ETI(NA,G.704)$_{5592}$

2. Specification of an ensemble configuration consisting of one EPP and several SPPs

   a) Specification of the EPP
      - allocation of ensemble identifier (EId) and ensemble label
      - stipulations for the FIC (FIC capacity and FIGs allowed to send)
      - specification of a FIG file including FIGs for signalling of ensemble information
   b) Specification of SPPs for all service providers
      - allocation of service provider identifier (SPId)
      - allocation of resources (MCI and FIC capacities, coding ranges, streams, etc.)
      - stipulations for the FIC (FIGs allowed to send, announcement parameters)
      - specification of physical input;

   c) Additional specification of MCI and SI (for non-STI service providers only)

3. Manual or scheduled download of the ensemble configuration specified (including valid time).

The DM001/STI will then generate the ETI stream according to the specifications downloaded. A reconfiguration will be carried out and the STI-C(TA) control connections to the STI service providers will be opened. After that, the STI service providers are able to define their virtual ensemble and to initiate reconfigurations of their own.

Monitoring of the ensemble multiplexer can be done by means of the status windows at the ensemble controller. Additionally a log file will be generated to store messages about all relevant events. A message browser for filtered evaluation of the log file is also provided.

Figure 6.18 displays the conditions at the service provider side. The MAGIC/STI system provides not only sufficient audio encoding functionality but also realisation of service multiplex, that is several services or service components can be handled.

MAGIC/STI basic units can be expanded by up to seven further encoders, if several audio signals are to be encoded at the same location. Encoders are also able to insert PAD into the audio data stream. Audio signals are fed either as analogue signals or as digital AES/EBU signals.

The output interface complies with G.703/G.704 (2 Mbit/s). Additionally data services in packet mode format can be inserted via the STI-D input. Running under WindowsNT,the service controller software consists of three major modules:

- Main manager
- Control file manager
- FIG encoder.

The main manager controls in a central function collaboration with the MAGIC/STI audio encoder(s) and service multiplexer. It accepts inputs from the service provider operator and serves to define and administrate the streams to be handled. It is also used to specify scheduled actions, to establish event-triggered signalling and co-ordinates collaboration with the control file manager and FIG encoder.

In its database the main manager stores all information relating to configurations and schedules. The control file manager complements the main manager in specifying and monitoring the respective MCI configuration, the specification of the virtual ensemble. Based on the SPP the service provider operator is able to specify sub-channel and service organisation, that is assignment of bit rates, start addresses, protection levels and so on. Consistency of the defined configuration is checked immediately.

The task of the FIG encoder is to provide FIG data for insertion into the FIC of the DAB signal. Since FIC data originate at the service provider and at the ensemble provider as well, the FIG encoder can be used by both sides equally. At the service provider side the FIG encoder supports the encoding of static and quasi-static FIG data. Resulting FIGs are embedded within FIG files and transported via STI-C to the ensemble provider in order to be stored there and passed into the ETI data stream if activated. Conversely, FIG data from the ensemble multiplexer database can be recalled and depicted for monitoring purposes. In order to initiate signalling via the FIG file mechanism, the service provider software can be coupled to the studio control process. For instance, announcement switching can be event triggered accordingly.

Furthermore, precautions have been implemented to adapt the whole network solution to a separate quality supervision system of the DAB network operator.

6.4.4   Experience and Outlook

The stepwise-developed broadcast network solution was thoroughly tested in the laboratory. After this, a field test within L-band DAB net Berlin was carried out successfully. Two broadcasters were selected to play the role of STI service providers, while the ensemble multiplexer, encompassing four further service providers, was operated by Deutsche Telekom.

The field test has provided evidence that the implementation fulfils the requirements of reliable handling of the services and supporting signalling within the collection and distribution network. For instance, the successful operation of traffic announcements as a necessary feature in DAB as it is in FM/RDS has been an important result. With respect to the existing infrastructure of the STI broadcasters, two possibilities of event-oriented FIG file activation control were applied: manual triggering of announcements from the programme moderator's button, as well as announcement triggering by derived signal from the studio operating system. Of course, the overall signal delay of the DAB chain has to be taken into account by the speaker, that is the speaker cannot start an announcement immediately after actuation of the procedure.

Also, linking functionality worked well by application of STI in both cases:

•   explicit signalling with service linking FIG 0/6
•   implicit signalling based on identical codes for FM/RDS PI code and SId of DAB.

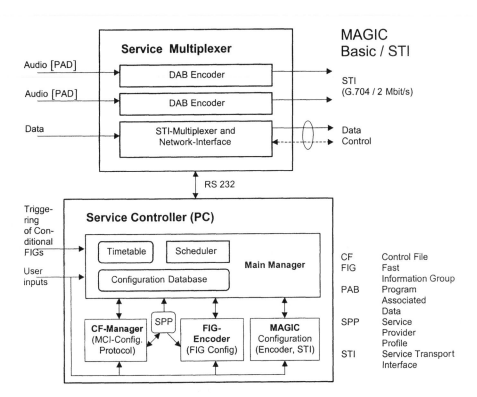

Figure 6.18   STI solution at service provider

The developed network solution corresponds to STI Level 3 according to (TRS 101860). This offers the chance to support regular DAB operation at a good level, and actually to exploit the superior system features of DAB in comparison to FM/RDS. In future it will be important to make further progress concerning enhanced management (see section 6.2.6) as well as provision of open and interoperable system solutions.

7

The Broadcast Side

WOLFRAM TITZE and STEPHEN BAILY

7.1   General

This chapter explains the broadcast side of DAB, namely the transmitters and the corresponding transmission networks. Planning and evaluation aspects for single frequency networks are also covered. Signal delivery to transmitter sites is covered in detail in Chapter 6.

7.1.1   Difference between FM and DAB Networks

Planning of transmission networks for FM broadcasting is traditionally based on the concept of multiple frequency networks (MFNs). In an MFN, adjacent transmitters radiate the same programme but operate on different frequencies to avoid interference of the signals where the coverage areas of different transmitters overlap. Basic FM receivers cannot cope with interfering signals from other transmitters of the same network. Coverage planning for an FM network requires frequency planning for the different transmitter sites, to optimise use of the scarce resource "frequency".

DAB in contrast allows single frequency networks (SFNs), where all transmitters of the network transmit exactly the same information on the same frequency. The main condition for a working SFN is that all transmitters are synchronised to each other in frequency and fulfil certain time delay requirements which will be explained later in this chapter. Coverage planning for a DAB network requires time delay planning between the different transmitters instead of frequency planning as in the case of FM. The SFN capability of DAB allows complete coverage of very large regions without the receiver having to tune to a different frequency while moving around in the area.

Digital Audio Broadcasting: Principles and Applications, edited by W. Hoeg and T. Lauterbach
©2001 John Wiley & Sons, Ltd.

In contrast to FM broadcasting, DAB transmits typically five to seven different programmes in one single ensemble on one frequency and all programmes contained in that multiplex share the same coverage area. Distinction by coverage area is therefore not possible for radio stations whose programmes share the same multiplex. It is also not advisable in an SFN to introduce local windows, that is areas where some transmitters of the SFN radiate a slightly different multiplex to achieve local programme variation. By definition, local windows cause problems for the receiver in the overlap area of the differing programmes of the multiplex since it cannot determine which programme to select.

7.1.2   Why SFNs Are Possible with DAB

DAB is a digital broadcasting system which was especially developed for the challenging transmission characteristics of the mobile radio channel. Typical phenomena of this channel like Doppler shift and multipath propagation with the resulting time and frequency selective fading had to be taken into account while developing DAB. To cope with these problems, the guard interval was introduced between consecutive data symbols, time and frequency interleaving techniques were applied to the data stream, a choice of sub-carrier spacings in the multicarrier modulation scheme was introduced and channel coding techniques were applied to correct for transmission errors. Table 7.1 shows the most important problems of mobile radio transmission systems and how they are solved in DAB.

Table 7.1 Problems of mobile radio transmission systems and their solution in DAB

Problems	DAB Solution
Time-dependent fading (multipath while driving)	Time interleaving
Frequency-dependent fading (stationary multipath)	Broadband system with frequency interleaving
Doppler spread (speed dependent, while driving)	Sub-carrier spacing as a function of the transmission frequency
Delay spread (due to multipath)	Guard interval (allows SFNs)
Transmission errors	RCPC (Rate Compatible Punctured Convolutional) codes to reconstruct the original bit stream

Four different transmission modes were developed to cater for a wide range of speed and frequency requirements in DAB systems as shown in Figure 7.1. A detailed explanation of the mobile radio channel and the DAB transmission scheme can be found in Chapter 2.

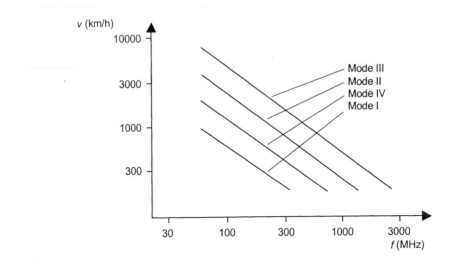

Figure 7.1 Interdependencies of maximum speed, frequency and DAB mode

The DAB system is a very robust and frequency economical transmission system which enables correct decoding of information despite Doppler spread and multipath reception. The effect of multipath reception is depicted graphically in Figure 7.2. Both the direct signal from the transmitter and reflected signals arrive at the antenna of a receiver.

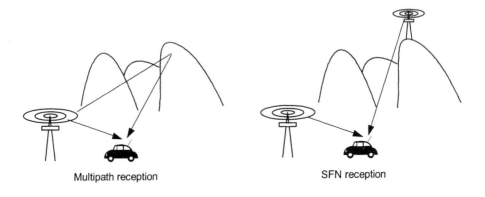

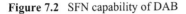

Multipath reception   SFN reception

Figure 7.2 SFN capability of DAB

All signals contain identical information but arrive at different instances of time. Owing to the measures previously described, the DAB receiver is able to cope with these multipath signals.

Going a step further, it is now irrelevant for the receiver whether the delayed signals were originated by the same transmitter or come from another transmitter that transmits exactly the same information synchronised in time as shown in the second part of Figure 7.2. This means that DAB allows coverage of any area with a number of transmitters that transmit the identical programme on the same frequency. Such broadcast networks are called SFNs.

The SFN capability of DAB transmission networks can be seen as an extra benefit which is an implicit result of the initial requirement during the development of DAB to cope with multipath phenomena typical of mobile radio reception.

7.1.3   Advantages of SFNs

In the following section two main advantages of SFNs over MFNs will be briefly introduced: power economy and frequency economy. A more detailed discussion of these issues can be found in section 7.3.

7.1.3.1   Power Economy of SFNs

Owing to the properties of the system, DAB receivers can use all signals received in an SFN in a constructive manner. This works as long as all signals arrive within the guard interval. Signals with longer delays create self-interference problems in the SFN and must be avoided by careful network planning.

Moreover, SFNs also show a diversity effect due to transmitters at different locations in the network. The diversity effect, that is the fact that the probability of simultaneous shadowing in the presence of several signals is much lower than the probability for shadowing for a single signal, results in an additional statistical network gain.

DAB networks are therefore very power economical. The advantages offered by SFN operation, and the properties of the digital transmission system itself, allow lower transmitter powers compared to FM for the same coverage quality. Power savings can be as high as 10 dB. Figure 7.3 shows this effect in a qualitative manner. To cover the same area with one programme on the same frequency, FM needs one high-power transmitter, whereas DAB with an SFN network with several transmitters needs much lower transmitter power in total. Another effect of the DAB SFN compared to FM networks is the much lower spill-over which causes unwanted interference in neighbouring networks.

7.1.3.2   Frequency Economy of SFNs

The fact that SFNs allow coverage of large areas occupying only a single frequency in the spectrum results in highly frequency economical planning possibilities for complex broadcast landscapes. The SFN technology also allows successive improvement of coverage quality without having to replan frequency allocations. Coverage problems within the network can be solved by simply putting up additional transmitters. The additional planning that must be done is to check the timing constraints of the SFN to avoid violation of the timing budget given by the guard interval of the chosen DAB transmission mode. Note that extra care must be taken to avoid an impact on reception in distant parts of the network because of spillover from the additional transmitter during abnormal propagation conditions.

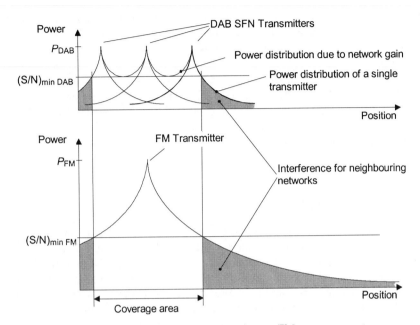

Figure 7.3   Power economy of DAB SFNs in comparison to FM

7.1.4   Layout of the Rest of the Chapter

Section 7.1 gave a brief introduction to the topic of DAB broadcast networks without touching upon the issue of signal delivery to the different transmission sites of the network. This aspect is covered in detail in Chapter 6. As explained so far, the major difference of DAB networks compared to conventional FM networks is the SFN capability of DAB. The rest of this chapter will therefore revolve around the topic of SFNs. Section 7.2 describes the equipment needed on the transmitter site to set up SFNs. Section 7.3 deals with the concept of SFNs in detail. Section 7.4 explains which aspects must be considered when planning networks with SFN coverage. In section 7.5 the issues of coverage evaluation and monitoring of SFNs are discussed and section 7.6 covers the aspect of frequency management and frequency allocation for DAB networks.

7.2   DAB Transmitter

7.2.1   General Aspects

Figure 7.4 shows the block diagram of a DAB transmitter. Each transmitter consists of a number of functional blocks which will now be explained. The ETI output signal from the ensemble multiplexer is delivered to the transmitter site via the DAB distribution network. At the input of the transmitter the signal is delayed to synchronise the SFN in time. After COFDM encoding the baseband output signal of a COFDM encoder can be subjected to further signal processing for non-linear pre-distortion or crest factor manipulation before it is converted from digital to analogue. After conversion to the analogue domain the signal is upconverted to the desired final radio frequency. Finally the RF signal is amplified and filtered to fulfil the relevant spectrum masks before it is radiated.

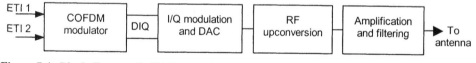

Figure 7.4   Block diagram of a DAB transmitter

7.2.2   Signal Processing Blocks of a COFDM Modulator

COFDM modulators usually contain not just the pure DAB signal processing part but also an input stage to process the different variants of the ETI signal and to perform signal delay. The output signal of the modulator is either the DIQ (Digital In-phase and Quadrature) baseband signal according to (EN 300 798) or an RF signal at a convenient IF or RF if an I/Q modulator is included. The signal processing blocks of a COFDM modulator are shown in Figure 7.5 and described in the following paragraphs.

The input stage strips the ETI signal (both ETI(NI) and ETI(NA), see Chapter 6) down to the ETI(LI) level as described in (EN 300 799), the logical interface level. Since the ETI signal is transmitted in HDB3 format, it is also converted to TTL level in the input stage. If the modulator has two inputs to allow networks with redundant distribution paths, both inputs are monitored for signal quality in this stage of the modulator and selection of one of the two input signals for further processing is carried out here.

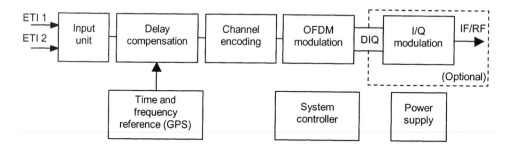

Figure 7.5   Block diagram of a COFDM modulator

In the delay compensation section the input signal is delayed within a range of zero to sometimes more than 1 second, typically in steps of 488 ns (488 ns are conveniently available in the system since they are the period of the ETI signal which is delivered at a rate of 2,048 Mbit/s). For dynamic delay compensation, the time-stamp of the ETI(NI) or the ETI(NA) is evaluated using the 1 pps (pulse per second) signal from a GPS (Global Positioning System) receiver as time reference. The information for an automatic static delay compensation is transmitted in the MNSC (Multiplex Network Signalling Channel) of the ETI. This delay is called transmitter offset delay and can be set for each transmitter individually making use of the unique encoder Id. The user can also set a separately adjustable delay for each input which is referred to as manual delay compensation. Combinations of the different methods of delay compensation are possible. A detailed treatment of the different delays in a DAB network is given in section 7.4.3.

The channel coding block performs all the encoding necessary to achieve a high level of signal robustness and to allow error correction in the case of bad transmission. Energy dispersal, convolutional encoding, MSC (Main Service Channel) time interleaving, MSC multiplexing, transmission frame multiplexing and frequency interleaving according to (EN 300 401) are performed in this block.

In the OFDM modulation block, the output bit stream from the channel coding block is mapped on to DQPSK (Differential Quadrature Phase Shift Keying) symbols, before $x/\sin x$ pre-correction for the digital-to-analogue conversion is performed. Finally the IFFT (Inverse Fast Fourier Transformation) with generation of phase reference symbol, TII (Transmitter Identification Information) and guard interval are performed to generate the DIQ baseband signal.

The DIQ baseband signal can be used to perform further signal processing like non-linear pre-correction or crest factor manipulation. To complete the COFDM modulator, a system controller and power supply are needed and an I/Q modulator may be incorporated.

7.2.3   Requirements of Analogue Processing

Analogue processing in the DAB transmitter must convert the digital representation of the COFDM signal to its final frequency and power, while keeping the distortion of the signal within acceptable limits.

Intermodulation is normally the dominant distortion to be considered in a DAB transmitter, and applies to all parts of the analogue processing, but there are a number of other factors that also need to be considered. These include frequency response (both amplitude and phase effects) of the signal path, and oscillator phase noise.

7.2.4   Digital-to-analogue Conversion

The output from the digital processing, and thus the input to analogue processing in the transmitter, is nominally the DIQ signal defined in (EN 300 798). This standard specifies a resolution for the I and Q signals of 8 bits, and a "clip-to-RMS" ratio for these signals of 12 dB. This specification is intended to allow digital and analogue sections from different manufacturers to work together. However, many manufacturers supply both sections, and internally may use slightly different settings (e.g. higher resolution to reduce the quantising noise floor, or a different clip-to-RMS ratio).

The example transmitter in Figure 7.6 shows individual baseband I and Q chains, each consisting of DAC and baseband filtering, followed by upconversion to a final frequency, amplification, and band-pass filtering.

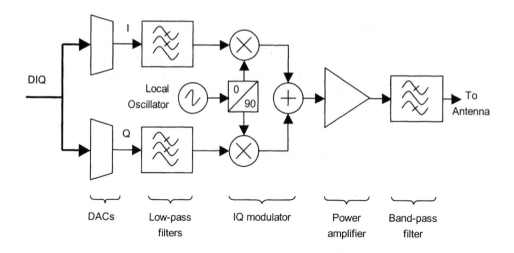

Figure 7.6 Example block diagram of a DAB transmitter with DIQ baseband input

In this arrangement it is important that DACs (digital-to-analogue converters), filtering and upconversion paths for I and Q signals are very closely matched, in amplitude and phase, across the bandwidth of the baseband I and Q signals. Any imbalance will result in distortion of the final RF signal, resulting in additional in-band components that will raise the in-band noise floor.

Further considerations include the phase accuracy of the in-phase and quadrature local oscillator signals, and the linearity of the upconversion mixers. Low-level breakthrough of the local oscillator, at the centre frequency of a DAB signal, can be tolerated because this carrier position is not modulated.

Most early implementations of DAB transmitters used this approach, and with careful alignment very good results can be achieved. One of the advantages of the approach is that transmitters at Band III can be implemented by a single upconversion to final frequency, avoiding the need for IFs requiring filtering and further upconversion.

An alternative technique, in which the upconversion to a low IF (normally a few MHz) is performed in the digital domain, is illustrated in Figure 7.7. This is also common and avoids the need for closely matched DACs and filters. In addition, highly accurate local oscillator (LO) quadrature can be achieved, and LO breakthrough eliminated (although this is not a major problem for DAB). However, further upconversion is required in this case.

One further factor to be taken into account is the spectral purity of the oscillators used in the transmitter, most commonly expressed as single-sideband phase noise (or "phase noise" for short). OFDM systems are tolerant of phase noise to some extent, but beyond certain limits suffer from adverse effects.

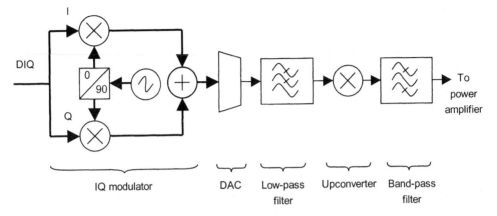

Figure 7.7 Block diagram of DAB transmitter using low IF

The first effect is mainly due to oscillator phase noise at frequency offsets equal to, or greater than, the carrier spacing of the OFDM system. This contribution to the phase noise causes leakage of the carriers into one another. The carriers are no longer genuinely orthogonal, and cause a small amount of interference to one another. This phenomenon is therefore often termed "intercarrier interference".

The second effect is mainly due to oscillator phase noise at frequency offsets less than the carrier spacing. Because the data are carried by changes in phase between successive symbols, the phase noise causes a reduction in the inherent signal-to-noise ratio. This effect is often termed "common phase error", because it affects all carriers equally.

The effects of oscillator phase noise in OFDM systems can be analysed in detail for individual oscillator characteristics using weighting functions (e.g. Stott, 1996), but for practical purposes, common phase error is normally the dominant effect in DAB systems. Good engineering practice requires that the degradation of signal-to-noise due to the transmitter from these effects is minimised, and a common rule of thumb is that the phase noise should be no more than -60 dBc/Hz at an offset of 25% of the carrier spacing (TR 101496).

7.2.5   RF Upconversion

As noted above, upconversion is required if the in-phase and quadrature signals are modulated on to an IF rather than the final RF. This applies for some Band III transmitter implementations, and most if not all L-band implementations. Most of the considerations discussed above apply equally to the upconversion process, including mixer linearity, spectral purity of oscillators, and frequency response of band-pass filters used to remove LO breakthrough and image products.

7.2.6   *Amplification and Filtering*

Because COFDM is a multicarrier system, an important cause of signal distortion is amplifier non-linearity. This causes intermodulation between the carriers, giving rise to RF products both within the bandwidth of the COFDM signal itself, and also in adjacent channels (Figure 7.8).

The difference between the spectral density of the COFDM signal and the out-of-band intermodulation products is often referred to as "shoulder height" (in Figure 7.8, for example, the signal distortion due to the non-linearities of the power amplifier has resulted in a shoulder height of 26 dB). Control of out-of-band products is important, because their level affects the performance of other signals in the adjacent channel.

The in-band intermodulation products are hidden by the COFDM signal itself, but their level is typically around 3dB higher than the shoulder height of the out-of-band products immediately adjacent to the ensemble. These products limit the performance of the DAB transmission, because they represent a "noise floor" that cannot be removed in the receiver. Accordingly, transmitters are designed to ensure that the in-band "floor" is sufficiently low that its impact on system performance is minimal.

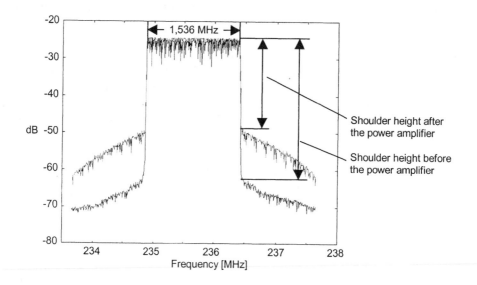

Figure 7.8 Spectrum of DAB signal with amplifier non-linearity

Having chosen the acceptable level of in-band "floor", a similar level of out-of-band products will be generated. However, to ensure efficient use of spectrum, (EN 300 401) and other standards require a much lower level of out-of-band emission by the transmitter, and this can be achieved only by external band-pass filtering. The attenuation requirements of the filter depend on the level of out-of-band emissions at the output of the amplifier, and therefore on the linearity of the amplifier and preceding stages. As a result, one of the major design decisions in a DAB transmitter is the trade-off between filter performance and amplifier linearity.

The most efficient amplifiers tend to be highly non-linear, and are not suitable for DAB. Conversely, highly linear amplifiers tend to be inefficient. Most DAB transmitters are designed so that the final power amplifier is the dominant non-linearity in the system, which allows it to be as efficient at possible. Even so, this amplifier requires a "back-off" of several dB (i.e. the output power of the DAB signal is several dB below the saturated output power of the amplifier when passing an unmodulated RF signal). For solid state amplifiers, the back-off is typically between 6 and 8 dB.

Operating amplifiers in this fashion results in a shoulder height of 25 to 30 dB. In order to meet the out-of-band spectrum requirements of (EN 300 401), high-order passband filters are required (e.g. in Band III eighth-order filters are commonly used). These filters are normally the dominant cause of frequency selective effects in the DAB transmitter, in terms of both amplitude and phase (although baseband filters can also have an influence). Phase effects in particular need to be kept within acceptable limits. If this is not done, the performance of the DAB system can be degraded (e.g. a large spread of group delay across the ensemble can reduce the useful guard interval at the receiver).

Owing to the expense of high-power filters and amplifiers, recent years have seen much interest in improving the efficiency of DAB amplifiers without adversely affecting the spectrum. Analogue pre-correction techniques, in use in broadcast transmitters for many years, are now being replaced by more sophisticated techniques such as closed-loop systems and various pre-conditioning or adaptive pre-correction methods implemented largely in the digital baseband domain. One of the techniques is to apply non-linear pre-correction to the signal to linearise the non-linear power amplifiers. As a result the in-band "noise floor" and the out-of-band emissions are reduced, the latter giving much better shoulder heights of the output spectrum. Another technique is crest factor manipulation which can be used to achieve significantly higher output power of the transmitter. However, crest factor manipulation degrades the overall system performance due to additional signal distortion. It should only be applied in combination with non-linear pre-correction techniques since they improve the overall system performance.

7.3   SFNs

7.3.1   Review of COFDM Principles

As stated earlier in the chapter, COFDM was originally intended to provide successful reception in multipath propagation conditions arising from reflected signals, but it works equally well for reception of multiple transmitters carrying the same signals. This gives rise to the possibility of an SFN, in which all transmitters carry the same information at the same time (or nearly the same time). Important factors for successful implementation of SFNs are the accuracy of the frequency and the timing of each transmitter. In addition, the length of the guard interval is important in SFN implementation, because it influences the transmitter spacing.

7.3.2   Time and Frequency Synchronisation

For an SFN to operate effectively, the transmitters must deliver the DAB signal to the receiver at the same time, or nearly the same time, and at the same frequency.

Frequency errors between the transmitters cause a loss of orthogonality between the received carriers, and also reduce the receiver's tolerance to the Doppler effects experienced in mobile reception.

Timing errors between transmitters erode the guard interval of the composite received signal, and can therefore disrupt the performance of the SFN.

For these reasons, transmitter networks are normally specified to keep transmitter frequencies within 1% or so of the carrier spacing, and timing within a few per cent of the guard interval. In order to achieve this, an independent and ubiquitous time and frequency reference is required. The GPS is commonly used for this purpose. GPS receivers are available that offer time and frequency references, typically 1 pulse per second (pps) and 10 MHz signals, with accuracy well in excess of that required for DAB SFNs. The 1 pps signal is used to define the transmission time of the data, and the 10 MHz signal is used as a reference for LO synthesisers that determine the final radio frequency.

7.3.3   Signal Reinforcement from Nearby Transmitters, Network Gain

If two or more transmitters serve the same area, their signal strengths are, in general, not strongly correlated. The signal strength of the transmitters varies with location, but because the signals are not strongly correlated, an area of low signal strength from one transmitter may be "filled" by a higher signal strength from another transmitter. In RDS (Radio Data System, used in FM networks to carry additional information), this is exploited by allowing the receiver to retune to alternative frequencies carrying the same programme if reception of a particular frequency is poor; that is, there is frequency diversity. However, unless two receiver front-ends are used, the receiver has to retune without prior knowledge of whether the alternative frequencies will offer greater signal strength.

In a DAB SFN, retuning is not necessary. For multiple transmitters on the same frequency, it remains true that an area of low signal strength from one transmitter may be "filled" by a higher signal strength from another transmitter. This is a form of "on-frequency" diversity, in which the receiver does not have to retune, although it may adjust its synchronisation to make best use of the available signals.

Another way of looking at this phenomenon is to consider the aggregate strength of the composite signal from a number of transmitters. This varies less with location than the signal strength from any of the individual transmitters, or in statistical terms, the variance of the strength of the composite signal is lower.

This effect offers a major advantage. In a multifrequency network, a number of transmitters may provide signal strength to an area without providing adequate coverage. This is less common in SFNs because of the tendency of two or more transmitters to fill each other's coverage deficiencies. This results in the coverage of an SFN being greater than the sum of the coverages of its individual transmitters, and is often known as "network gain".

7.3.4   Effects of Distant Transmitters

Nearby transmitters on the same frequency have a constructive effect, but in a large SFN, the more distant transmitters, whose signals may arrive outside the guard interval, can act as interferers. Although it is possible for signals arriving just beyond the guard interval to contribute some useful energy, this is a complicated issue influenced by the design of the receiver, and therefore the effects are somewhat variable.

Transmitter spacing is therefore a factor in network design, but in practice the availability of suitable transmitter sites, topography and population density are also major influences. For Band III SFNs, using Mode I, transmitter spacings are usually somewhat lower than the distance corresponding to the guard interval (around 75 km). This may result in many transmitters contributing useful energy under favourable circumstances, but in SFNs above a certain size there will always be potential for interference from distant transmitters. Because of this, SFNs have to be planned taking very careful account of the interference to distant locations caused by transmitters, as well as the coverage provided in their immediate surroundings.

7.3.5   Optimised Coverage of SFNs

Although the transmitters in an SFN need to deliver the signals with very precise timing, it is not always necessary for them to be exactly co-timed, and in some circumstances it can be advantageous to offset the timing of particular transmitters by significant fractions of the guard interval. This is particularly true at the extremities of coverage, or where low-power transmitters are used to fill gaps within coverage provided primarily by high-power transmitters. Transmitter timing is a variable in network design and can be used in combination with transmitter powers and radiation patterns to optimise the coverage of the network.

7.3.6   Closure of Coverage Gaps Using Gap Fillers

In conventional MFNs as used for FM broadcasting, coverage gaps are closed with additional transmitters, which need their individual frequency assigned. This requires careful planning to ensure that incoming and outgoing interference are correctly managed, and is often not economical with the frequency spectrum.

SFNs, however, allow relatively simple filling of areas not well served by the main transmitters, that is gaps in the coverage area, by installing additional retransmitters which operate on the same frequency as the rest of the SFN (see Figure 7.9). These additional retransmitters with a typical output power of the order of a few watts are called gap fillers or repeaters. Gap fillers must be located at points in the network where there is sufficient incoming field strength, normally received with a highly directional antenna, and where the transmitting antenna can fire into the as yet uncovered area of the SFN.

In the Canadian DAB networks more powerful retransmitters with an output power of up to 100 W are also used. This type of device is called a coverage extender since it is typically located at the fringe of the network and fires beyond the coverage area of a main transmitter. Coverage extenders enlarge the total area of the SFN instead of serving regions not well covered owing to local shadowing as gap fillers do. Regarding planning and synchronisation aspects, the same rules apply to both sets of devices.

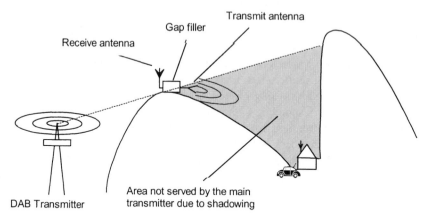

Figure 7.9 Typical application of a DAB gap filler

Gap fillers do not have to be exactly time synchronised to the other transmitters in the SFN. The gap filler simply amplifies the received DAB signal. To achieve good performance the received signal may be downconverted to an IF or even baseband for signal conditioning before it is upconverted again for final amplification and filtering. De- and remodulation for signal improvement are not done in gap fillers because the processing delay inherent in the DAB system means that the retransmitted signal would lie outside the guard interval irrespective of the DAB mode chosen. When a gap filler is installed, it must be ensured that the transmit and receive antennas are sufficiently decoupled to avoid unwanted feedback and blocking effects.

In areas where several DAB blocks are in use, the input stage of the gap filler must be block selective to guarantee that only the wanted signal is amplified. It should also be noted that in areas with gap fillers in a DAB network, geographical position estimation of the receiver using the TII feature will not be possible for two reasons. Firstly, because the transmitter time delay signalled in FIG 0/22 (TII field) is only valid for directly received signals and not for signals which suffer from the additional delay due to the signal processing in a gap filler, and secondly, if the signal from the main transmitter and one or more gap fillers is simultaneously received, distinction of the different signal sources will no longer be possible since they will all have the same TII identification. (Gap fillers cannot change the TII information of the signal.) For more information on TII in DAB see section 7.3.7.

7.3.7   Application of the TII Feature in SFNs

DAB allows identification of individual transmitters in an SFN with the TII feature. The TII signal is transmitted every other null symbol to allow the receiver to perform channel state analysis in null symbols without the TII signal. The TII signal consists of a certain number of pairs of adjacent carriers of an OFDM symbol and the actual pattern of the carrier pairs identifies the individual transmitter.

The identification of each transmitter is given by two parameters: the pattern and comb number, also called the main and sub-identifier of a transmitter. FIG 0/22 in the FIC of the DAB signal describes a set of parameters, the TII field, which contains all information necessary for the unique description of a transmitter. These parameters are transmitter identifiers, geographical location of the transmitter and the time offset of the transmitter (see section 7.4.3).

The main identifier is used to describe a cluster of transmitters in a certain region and each transmitter within a cluster has its own sub-identifier. Table 7.2 gives the number of possible main- and sub-identifiers as a function of the DAB mode.

Table 7.2 TII parameters for different DAB modes

Mode	Number of Main identifiers (= diff. patterns)	Number of Sub-identifiers (= diff. combs)	Number of Carrier Pairs Used per Comb
I, IV, II	70	24	4 of 8
III	6	24	2 of 4

Each comb number identifies a number of carrier pairs of which only half is used in a TII symbol. Which of the carrier pairs are used is determined by the associated pattern number. Since each comb number identifies a unique set of carrier pairs (i.e. each carrier pair is only used by a specific comb), the DAB receiver can simultaneously identify the signals of all transmitters with the same main identifier that is pattern number. To distinguish between transmitters with different main identifiers, the sub-identifier must be chosen carefully to avoid ambiguities. The exact relationship between comb and pattern is given in references (EN 300 401) and (TR 101 497).

The TII feature of DAB allows the receiver to calculate its position if signals from at least three transmitters are received and FIG 0/22 (TII field) is signalled in the FIC (Layer, 1998). The knowledge of the receiver position can be used for intelligent change of frequency when leaving the coverage area of the network. It can also aid the automatic selection of information, for example only information relevant for the current region is displayed.

7.4   Coverage Planning

7.4.1   Field Strength Considerations

Traditionally, coverage of analogue broadcast systems has been planned on the assumption that reception would make use of fixed antennas at roof height, nominally 10 m above ground level (AGL). For DAB, it was clear that this assumption would not be valid, for two main reasons.

The first reason is that a large proportion of radio listening takes place in a mobile environment, and therefore coverage would have to be planned for mobile receivers, with antennas at car roof height. For coverage planning purposes, this is nominally 1.5 m AGL. The reduction in field strength due to the lower antenna height has been studied and found to be typically 10 dB in Band III.

The second reason is that, as a digital system, DAB exhibits rapid failure characteristics, and the margin between a perfect signal and total failure is just a few dB. When deciding whether or not an area is "covered", it is necessary to ensure that the signal availability in that area is very high. For planning purposes, the required availability in a unit area is typically taken to be 99%. A common unit area used in coverage planning is a 1 km square, which can be said to be covered if the signal quality is adequate over 99% of the area of that square.

The signal quality required for adequate reception is determined by a number of factors, including the performance of the DAB system itself, receiver parameters, signal strength from "wanted" transmitters (i.e. those whose signals arrive within the guard interval of the receiver), and signal strength from any interfering transmitters.

When calculating link budgets for the DAB system (TR 101758), it is normal to assume an unfavourable reception environment, such as that in a mobile, fading environment. Under these circumstances the demodulator in a typical DAB receiver requires a signal-to-noise ratio of 14 dB. Making realistic assumptions regarding receiver noise figure and gain of a vehicle-mounted antenna (including any connecting leads), a value for the minimum field strength required for successful reception can be derived. In Band III, the required field strength is normally taken to be 37 dB(μV/m), although some references may give slightly different figures (within 1 or 2 dB).

The field strength of 37 dB(μV/m) is the minimum field strength required for reception at a single location, at car roof height. If an area such as a 1 km square is to be regarded as covered, the available field strength must exceed this minimum over 99% of the area. The variation of field strength with location has been modelled statistically and found to follow a distribution that is approximately log-normal, with a standard deviation of around 4 dB in suburban environments.

Typically, field strength planning models predict the median field strength, that is the value exceeded for 50% of locations, rather than 99% of locations. To convert from the 50% locations field strength to the 99% locations field strength, it is necessary to subtract approximately 9 dB (being 2.33 times the standard deviation of the log-normal field strength distribution). In addition, such planning models generally predict field strength at 10 m rather than 1.5 m AGL, as noted above, so an allowance of 10 dB has to be made for this.

Adding these two allowances together, it can be seen that a predicted median field strength of 56 dB(μV/m) at 10 m AGL corresponds to a 99% locations field strength of about 37 dB(μV/m) at 1.5 m AGL. Accordingly, in this example, a 1 km square is considered to be served if the predicted median field strength exceeds 56 dB(μV/m).

7.4.2   Interference Considerations

The previous section described the approach taken to predicting coverage on the basis of signal strength from nearby transmitters in a network. Strictly speaking, this applies only in the situation where noise is the factor limiting reception. In practice, interference from other networks, or distant transmitters within the same network, may be equally or more important, and it is necessary to plan services for adequate protection against such interference.

It is beyond the scope of this book to deal in detail with the propagation of broadcast signals, but in summary it can be said that the field strength due to distant transmitters can vary significantly with time of day and climatic conditions, among other factors. Because of this variation, interfering field strengths are often predicted using terrain-based statistical models. Interference from a number of distant transmitters may need to be taken into account, and this is normally built into coverage planning software.

Although terrain-based modelling is favoured for network planning, it is not always suitable, for example if detailed terrain data are not available, as may occur if significant interferers are located in neighbouring countries or overseas. Under these circumstances, other methods, such as ITU-R Rec. 370, are used (ITU-R 370, 1990).

Again taking the rapid failure characteristic of DAB into account, the criterion for coverage is that the wanted signal strength should be protected against interference for 99% of the time. Put another way, given an interfering field strength predicted to be present for 1% of the time, the wanted signal strength must exceed the level of the interferer by a sufficient margin to allow successful reception. The required signal-to-interference ratio is called the protection ratio.

7.4.3   Delay Considerations

As already mentioned in the introduction to this chapter, coverage planning in a DAB network means to a large extent delay planning. All transmitters in a DAB network must be time and frequency synchronised for proper operation of the SFN. It is common practice to use GPS receivers to provide a highly stable frequency and time reference at different points in the network, typically the ensemble multiplexer and transmitter sites. The transmitter output frequency is locked to the 10 MHz signal of the GPS receiver and the GPS 1 pps signal serves as a reference for the delay compensation at each transmitter.

Time synchronisation is performed by introducing an artificial delay in the distribution network of the SFN or at the different transmitter sites as shown in Figure 7.10. The total duration for which the signal must be delayed is the sum of four different types of delay (EN 300 799):

1. **Network compensation delay:** The time the ETI (Ensemble Transport Interface) output signal of the multiplexer is delayed on its way through the distribution network to the transmitter site is called the network path delay. To ensure that the overall delay to each of the transmitter sites is constant and of known value, a network compensation delay is added for each path through the distribution network. The largest network path delay to a certain transmitter site determines the network compensation delays necessary for all the other network paths. In a correctly adjusted SFN, the sum of network compensation and network path delay must be the same for each network path.

2. **Transmitter compensation delay:** The time the signal is delayed in the equipment at the transmitter site through channel encoding (i.e. COFDM modulation) and RF modulation is called the transmitter processing delay. This also includes delay due to

signal conditioning and RF processing in the amplifier and associated filtering. The processing delay of the equipment differs from manufacturer to manufacturer. To ensure that the overall delay between transmitter input and antenna output at each transmitter site is constant and of known value, a transmitter compensation delay is added individually for each transmitter in the network. The transmitter with the longest processing delay determines the transmitter compensation delay for all the other transmitters in the network. In a correctly adjusted SFN, the sum of transmitter compensation and transmitter processing delay must be the same for each transmitter in the network.

3. **Transmitter offset delay:** This delay can be set for each transmitter site individually and is necessary for network planning aspects to achieve optimised coverage in an SFN. Owing to the surrounding topology and differing effective output power, the coverage areas differ from transmitter to transmitter. The transmitter offset delay ensures that the signals from individual transmitters arrive approximately at the same time in the overlap zone of their coverage areas. The transmitter offset delay can be set individually for each transmitter from the ensemble multiplexer using the MNSC (Multiplex Network Signalling Channel) in the ETI stream.

4. **Network padding delay:** This delay is used by the network operator or the broadcaster to adjust the overall delay of the network. It allows equalisation of the overall delay of the DAB network with that of other networks such as FM or DVB. It can also be used to synchronise different DAB networks to allow radiation of co-timed synchronising symbols for speedier locking of DAB receivers in the case of tuning to a different frequency.

In summary, network and transmitter compensation delay ensure that the DAB signal is theoretically transmitted at the same point of time from each transmitter site in the network, whereas the transmitter offset delay takes care of the topology of the SFN. The padding delay is only relevant with respect to other broadcasting networks. The different types of delay and how they interrelate are given in Figure 7.10.

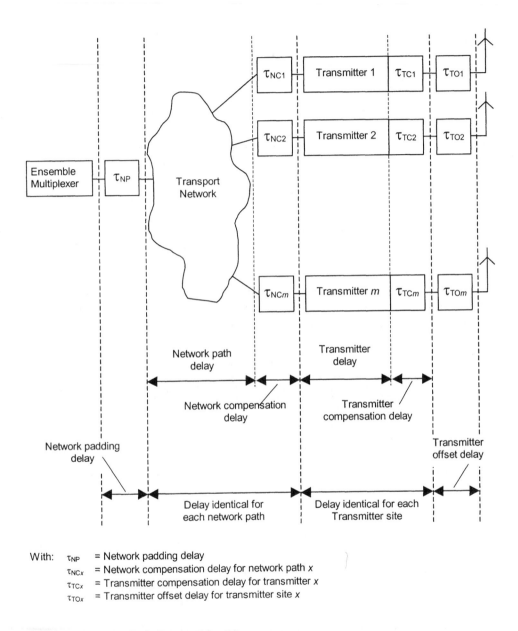

Figure 7.10 Types of delay in an SFN

In modern distribution networks, the network path delay may change as a result of the necessity to reroute the signal from time to time. To manage automatically this varying network path delay, dynamic delay compensation on the basis of time-stamps is used in DAB which allows synchronisation of the delivery of ETI frames at all transmitter sites. This requires a common time reference (e.g. GPS) at the ensemble multiplexer and all transmitter sites in the network. The time-stamp defines the instant of time at which the frame should be delivered to the channel encoder. The time-stamp can be carried in all variants of the ETI signal.

7.4.4   Detailed Planning

When planning networks in detail, the planner is subject to a number of constraints. These include the frequency range to be used, the receiving environment to be served, the availability of transmitter sites, and the allowable radiated power from these sites.

The frequency range will determine which of the possible DAB modes is most appropriate (see also Figure 7.1).

The choice of receiving environments is not limited to mobile reception, as described in section 7.4.1. Other receiving environments will impose different constraints; for example, indoor reception on portable sets will require additional field strength.

The availability of sites, and the allowable transmitted power, will have a strong influence on the design of the network and the distance between transmitters. As a rule of thumb it is usual to ensure that the spacing between adjacent transmitters is no more than the distance determined by the guard interval of the chosen mode, as set out in Table 7.3. However, in many circumstances a smaller spacing between transmitters is found to be necessary.

Table 7.3   DAB modes and approximate maximum transmitter spacing

DAB Mode	Guard interval (μs)	Approximate Maximum Transmitter Spacing (km)
I	246	74
II	62	18
III	31	9
IV	123	37

7.4.5   Example of an SFN in Band III

In the United Kingdom, the BBC's DAB network is an example of an SFN providing large-area coverage, and is illustrated in Figure 7.11. The network operates in Band III (Block 12B), and by Spring 1998 it comprised 27 transmitters, with radiated powers between 1 kW and 10 kW. This provided coverage to more than 60% of the UK population, and the network has since been enhanced with additional transmitters. In addition to achieving a high population coverage, the network was also designed to cover the major road routes between centres of population. Many major cities are served by more than one transmitter, so that the coverage would benefit from network gain.

The majority of these transmitters are co-timed, but there are exceptions. An example of this occurs in Northern Ireland, where the transmitter serving Belfast and the surrounding area (Divis) transmits with its signal advanced relative to other transmitters in the network. This timing means that the signal from Divis, arriving at the UK mainland, has a similar timing to that of the local transmitters, and is therefore a contributing transmitter in that area, rather than an interferer. This helps to improve the coverage of some densely populated parts of the north-west of England.

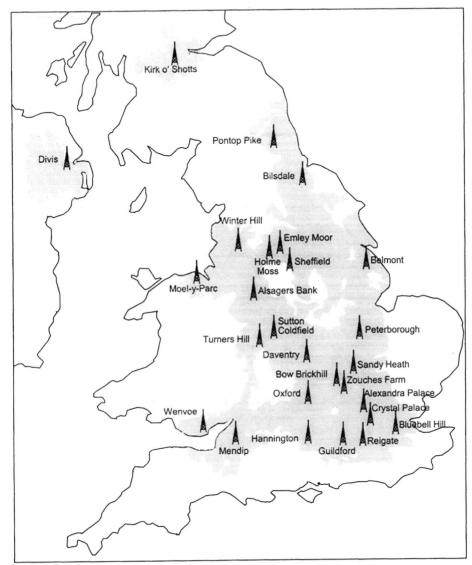

Figure 7.11   The BBC's SFN in the UK and Northern Ireland as of Spring 1998 [© Copyright British Broadcasting Corporation 2000. Reproduced by permission of the British Broadcasting Corporation]

7.4.6   Example of an SFN in L-band

Canada uses only L-band for its DAB networks and is therefore very well suited as an example for an SFN in this frequency range. In Ottawa, an L-band DAB network operating in Mode II with one main transmitter, one coverage extender and two gap fillers has been designed to serve the National Capital Region (Paiement, 1996; Paiement, 1997; Voyer 1996). This is an area of approximately 30 km in radius, comprising Ottawa and several smaller adjacent municipalities with a total population of close to one million people.

It was required to provide 99% service availability to the city core, the suburbs, portions of the main roads and highways surrounding the region. This objective is represented by a service polygon drawn as shaded contour over a simplified road map of the area (see Figure 7.12, covering an area of 92 km by 64 km). The predicted coverage achieved by the main transmitter is given in Figure 7.12 (availability at 99% of locations). It can be seen that there is a need for a second transmitter in the network to cover the south-west area of the specified service polygon.

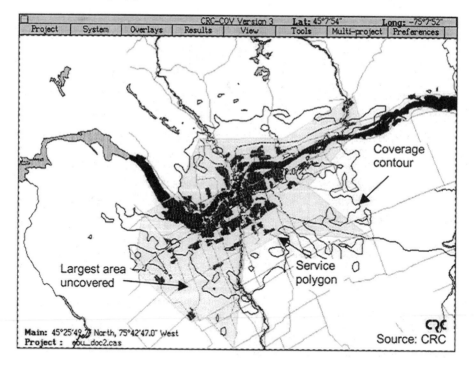

Figure 7.12 Service polygon for the Ottawa region and coverage of the main transmitter

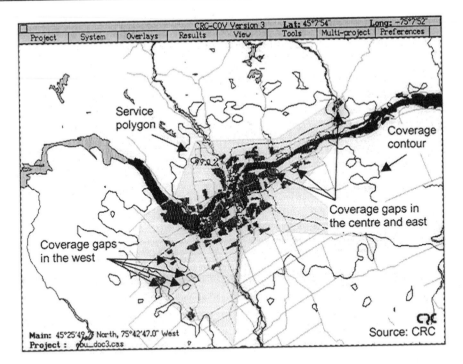

Figure 7.13   Coverage of main transmitter plus coverage extender

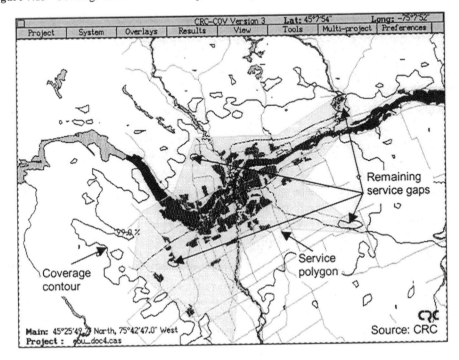

Figure 7.14   Coverage of main transmitter, coverage extender plus two gap fillers

The network planners then decided to use a coverage extender to enlarge the coverage area. The coverage predicted for the main transmitter plus coverage extender is given in Figure 7.13.

To close the coverage gaps in the west and east-centre of the required service area, two additional gap fillers were put into place. The complete predicted coverage of the Ottawa system is shown in Figure 7.14. The contour lines show four remaining small holes in the requested coverage area whereby three of them are not in densely populated areas.

The transmitter parameters are summarised in Table 7.4. It can be seen that the total service area of about 30 km in radius can be covered by four transmitter locations with a total amplifier output power of less than 200 W. This clearly demonstrates how power efficient DAB SFNs can be.

Table 7.4 Transmitter parameters of the Ottawa DAB system

	Main Transmitter	Coverage Extender	Gap Filler - West	Gap Filler – East-centre
ERP	1000 W	500 W	1000 W	1200 W
Amplifier power	100 W	50 W	30 W	12 W
Antenna gain	13 dBi	13 dBi	19 dBi	22.5 dBi
Antenna pattern	Omnidirectional	Omnidirectional	Directional, 90° beamwidth	Directional, 40° beamwidth
Antenna height	105 m	105 m	50 m	50 m
Distance from main transmitter	N/A	5.4 km	7.1 km	10.7 km

7.5   Coverage Evaluation and Monitoring of SFNs

7.5.1   Parameters for DAB Coverage Evaluation

In an FM MFN, field strength is the main parameter for evaluating the quality of the coverage. For DAB, however, more parameters are necessary to evaluate the quality of the SFN coverage. The following sections describe the main parameters and what they indicate:

1. **Field strength:** This parameter indicates whether there is signal coverage at all in this area of the SFN. However, sufficient field strength does not necessarily mean that proper reception of the DAB signal is possible. This signal level could be made out of contributions from several transmitters which are not properly synchronised in time and/or frequency.

2. **TII:** Analysis of the TII with a measurement system delivers the identity of the received transmitters, i.e. their pattern and comb numbers and the relative signal strength as shown in Figure 7.15. The corresponding FIG 0/22 gives additional information about transmitter location and time offset.

TII-Information		
ENSEMBLE	**LABEL**	**FREQUENCY**
00004258	DAB Bayern	223936 kHz

TRANSMITTER		**COMB**	**PATTERN**
1	49 %	4	1
2	30 %	7	2
3	21 %	6	2

Figure 7.15 Typical analysis of TII information

3. **Channel impulse response:** The channel impulse response shows in the time domain the different signal contributions which arrive at the receiver (see Figure 7.16). Evaluation of the channel impulse response shows whether the guard interval is violated or not. This parameter is therefore an indicator of the quality of the time synchronisation of the network. The channel impulse response together with the TII allows the identification of contributions from each transmitter. It must be noted that possible contributions from signal reflections due to the geographical topology must be taken into account in the evaluation process.

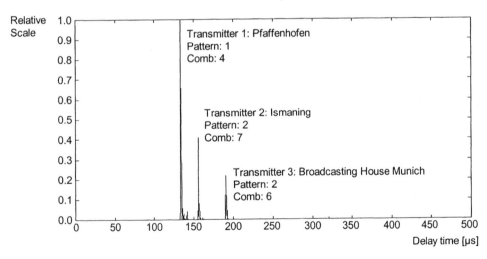

Figure 7.16 Typical plot of a channel impulse response

4. **Transmit frequency:** All transmitters in an SFN must operate exactly on the same frequency. To achieve a performance degradation of less than 1 dB, the minimum required accuracy of the output frequency is 10% of the sub-carrier spacing, that is 100 Hz in Mode I, 200 Hz in Mode IV, 400 Hz in Mode II and 800 Hz in Mode III (TR 101496). This parameter can be measured at the output of each transmitter in the SFN.

5. **Bit error rate (BER):** The BER is an indicator of the quality of the network. High BERs can be caused by not enough field strength or synchronisation errors in time and/or frequency. For good audio quality the BER should be 10^{-4} or better. The different types of BER are discussed in section 7.5.2. To indicate the quality of the received signal, CRC error measurements were discussed as alternatives to BER measurement at the early stages of DAB development. The DAB audio signal contains CRCs for the audio header and the scale factor which could be used. However, it turned out that these CRC errors are not sensitive enough since they only occur once the audio signal degrades audibly.

Coverage evaluation of a DAB network can be done by measuring and evaluating the above parameters. If the coverage is below the required level (i.e. a certain BER for typically 99% of the time at 99% of the area) careful analysis of this set of parameters and their interdependencies allows the planner to derive the measures necessary to improve the coverage quality.

All the parameters described so far represent well-measurable effects that do not vary much with time. However, to complete the coverage evaluation, investigation of long-distance interference must also be made. This type of interference depends heavily on propagation conditions and is only measurable for a small percentage of time. These effects are therefore usually modelled statistically in advance and taken care of in the planning phase of the network.

7.5.2   A Closer Look at BER Measurements

There are two different types of BER in a DAB system. Depending on the application, the channel BER or the net BER can be used as a measure for the quality of the coverage. In comparison, the net BER is the exact method for BER measurement whereas the channel BER stands for the pragmatic approach. The BER is defined as the ratio of number of errored bits received to total number of bits received. Strictly speaking this is the definition of a "bit error ratio" but it is commonly referred to as the "bit error rate".

7.5.2.1   Channel BER

The channel BER can be calculated by comparing the re-encoded bit sequence in the receiver to the actual bit sequence received. To measure the exact channel BER, the complete bit stream would have to be re-encoded in the receiver. An alternative to this concept is to re-encode only part of the received signal, that is only the first bit of the 4-bit mother code is re-encoded. Most receiver decoder chips contain this feature and generate a "pseudo-channel BER". Figure 7.17 gives the block diagram for the channel BER calculation using an MSC sub-channel as the reference bit stream. An alternative to the MSC BER is the FIC BER which uses the non-time-interleaved FIC data stream as the reference bit stream.

The shortcomings of this method can be clearly seen in Figure 7.17. The calculation of the (pseudo-)channel BER is only correct if the channel decoder is able to correct all transmission errors, i.e. it produces an ideal reference bit stream at the output. This is true for low bit errors and channel bit error measurement give good indications for BER in the order of 10^{-4} and lower. The advantage of this method lies in its simple implementation (Schramm, 1997).

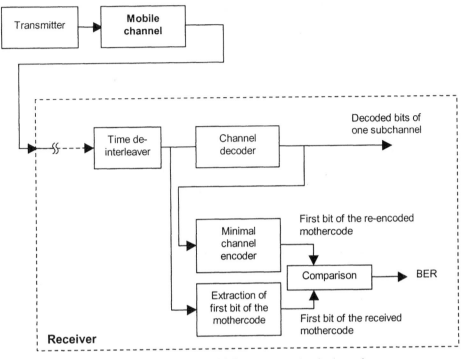

Figure 7.17 Generation of the pseudo-channel BER using a MSC sub-channel

7.5.2.2   Net BER

The net BER is calculated by comparing the received and decoded bit stream directly to the reference bit stream. In this case, the DAB transmission system is treated as a black box and a comparison of the output and the input signals of this black box give the net BER (Figure 7.18). To calculate the net BER, the reference signal must be known. For on-line measurements, a known test sequence must be transmitted, be it in a small sub-channel or in the PAD which takes up some of the capacity of the multiplex.

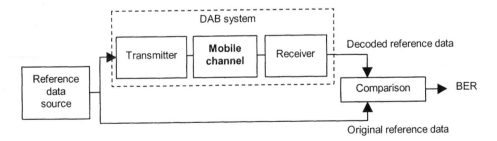

Figure 7.18   Generation of the net BER

An advantage of the net BER measurement is that the result is always correct irrespective of the quality of the transmission (Frieling, 1996). The net BER is interesting for service providers who are only interested in the quality of their service and treat the DAB system simply as another distribution channel for that service.

7.5.3   Timing Adjustment of SFNs

The only way to check for correct time synchronisation in an SFN is to evaluate the channel impulse response together with the TII. If the timing within the SFN is not correct, the delay times for certain transmitters must be adjusted. This can be done centrally in a DAB network from the ensemble multiplexer site. The ETI output signal of the multiplexer allows each transmitter site to be addressed and the delays of each transmitter to be set individually. This convenient feature avoids the time- and labour-consuming alternative of having to send people directly to the different transmitter sites to perform the necessary adjustments.

7.5.4   Monitoring of SFNs

Four parameters determine the quality of an SFN and must be constantly monitored: the correct transmit signal according to (EN 300 401), level of field strength, accuracy of transmit frequency and accuracy of timing between the different transmitters.

The correct transmit signal and the accuracy of transmit frequency can be monitored at each transmitter site using a reference receiver and a frequency counter. Monitoring the level of field strength must be done at two locations: firstly at the transmitter site by monitoring the output power of the transmitter and secondly in the field by constantly measuring the field strength and comparing it to the required values. If all is correct at the different transmitter sites, an alarm by a field strength measurement probe in the field would be triggered only by a change in propagation conditions. The timing of the SFN can only be monitored through units that measure the channel impulse response in the field. Changes of the channel impulse response can be caused by wrongly timed transmitters or a change in propagation conditions.

7.6   Frequency Management

7.6.1   General Aspects

In order to simplify transmitter and receiver design, agreement on the frequencies to be used for DAB was reached some years ago. All DAB transmissions have a centre frequency on a nominal 16 kHz lattice, and two frequency bands are, in effect, standardised for DAB, though their availability varies from country to country. These are Band III (174–240 MHz) and L-band (1452–1492 MHz). Within these bands, individual centre frequencies for DAB transmissions are specified, which simplifies management of the spectrum as well as receiver design. The spectrum occupied by an ensemble centred on a particular frequency is referred to as a "block". A full list of blocks and corresponding centre frequencies is given in (EN 50 248).

Band III presents significant problems of sharing with other services, including analogue television, private mobile radio and military users. Band III television imposes particular constraints in some countries, and for that reason the agreed DAB frequencies are specified to accommodate four DAB ensembles within a 7 MHz television channel. This is illustrated schematically for Channels 11 and 12 in Figure 7.19. The four blocks in each channel are numbered A to D.

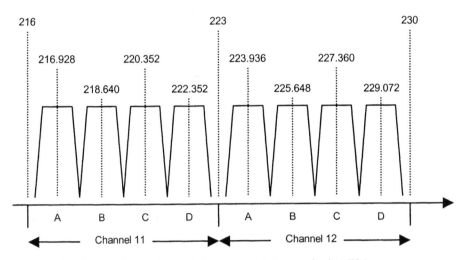

Figure 7.19 Block allocations in Channels 11 and 12 (all frequencies in MHz)

Owing to shortage of Band III spectrum, increased use of L-band for terrestrial-DAB is anticipated in the next few years, primarily for local services. In a number of countries such as France and Canada where Band III is not available at all, only the L-band spectrum can be used for terrestrial DAB (T-DAB). It should be noted that the upper part of the L-band spectrum is currently reserved for satellite DAB.

7.6.2   Allocation of Frequencies

International agreement is required before broadcasters can start to transmit services on a particular frequency or block. In order to reach such agreement, it is necessary to define criteria for sharing the use of frequency blocks in different geographical areas. For example, Block 12B is used for the BBC's national multiplex in the United Kingdom, but it is also used for smaller area coverage in mainland Europe, and it is important that these services should be able to co-exist without mutual interference.

The criteria for sharing are based around the predicted levels of interfering signals from one or more service areas into other areas using the same block. These signal levels must be kept below an acceptable threshold. Calculation of the levels is performed on the basis of specified service areas, and "reference networks" of transmitters that are felt to be sufficiently representative of a practical network. Signal levels arriving in an area, from the reference network in another co-channel area, are calculated, and if the levels are sufficiently low, then sharing of that block is possible. Planning models for the VHF and

L-band networks are given in the EBU document BPN003 (BPN003). Figure 7.20 shows an example of such a reference network. The main transmitter in the reference hexagon has 100 W output power and the six peripheral transmitters have 1 kW each. Since directional antennas are assumed for the peripheral transmitters, this set-up is called a closed hexagon structure. An effective antenna height of 150 m is assumed and the distance between each of the transmitters is 60 km for Band III.

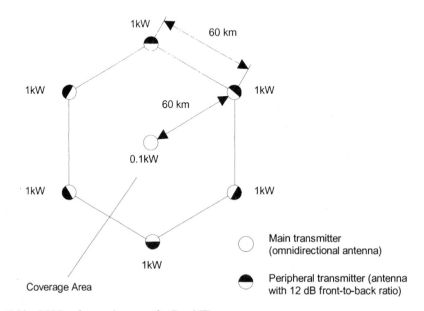

Figure 7.20 CEPT reference hexagon for Band III

The propagation model used in BPN003 is based on the ITU Recommendation 370 (ITU-R370, 1990) which assumes an ideal regular geometry of the coverage area, averaged channel characteristics and neglects topographical details. However, it is well suited for calculating the average potential for interference and the reuse distance of SFNs.

These techniques result in a frequency reuse pattern that is heavily dependent on human and physical geography, and the nature of broadcasting in individual countries. Figure 7.21 shows the part of the outcome of the CEPT Planning Conference held at Wiesbaden in 1995, which allocated two blocks ("priorities") to each geographical area in participating countries. This figure clearly illustrates how the different national approaches in Europe have resulted in markedly different uses of the same frequency, in this case Block 12B.

Once frequency allocations are agreed, real transmitter networks must be implemented. These will present a different set of interfering field strengths into neighbouring areas using the same frequency, but the general rule is that the sum of the interfering field strengths from the transmitters in the real network should not exceed that defined by the reference network. The Final Acts of the Wiesbaden Conference (CEPT, 1995) provide means of calculating the field strength summation.

In addition to multilateral international agreements, bilateral agreements between countries can allow additional use of frequency blocks, particularly where the geography is favourable. For example, in the United Kingdom, the Radiocommunications Agency has allocated seven blocks in Band III to T-DAB, and all of these blocks should soon be in use for national or local radio, although only two priorities were agreed at the Wiesbaden Conference. The use of the blocks is given in Table 7.5. The Radio Authority in the UK plans to use the frequencies available for local radio to license local multiplexes in 30 or more cities and regions, and at the time of writing this process is well under way.

Figure 7.21   Use of Block 12B (first priority) in Europe as agreed at the Wiesbaden
Conference in 1995 (shaded areas represent 12B coverage) [© Copyright
British Broadcasting Corporation 2000. Reproduced by permission of the
British Broadcasting Corporation]

Table 7.5 Block use in United Kingdom

Block	Status, Wiesbaden Conference	Used for
11B	None	Local multiplexes
11C	Second Priority (Isle of Man and Channel Isles)	Local multiplexes
11D	Second Priority (England and Wales)	Independent national multiplex (England and Wales); local multiplexes (Scotland)
12A	Second Priority (Scotland)	Independent national multiplex (Scotland); local multiplexes (England and Wales)
12B	First Priority (UK-wide)	Public service national multiplex
12C	None	Local multiplexes
12D	Second Priority (Northern Ireland)	Local multiplexes

8

The Receiving Side

TORSTEN MLASKO, MICHAEL BOLLE and DETLEF CLAWIN

8.1   General

DAB is different from the traditional analogue audio broadcasting systems like AM and FM. For example, DAB is a broadband transmission system, transmitting several audio programmes and data channels over the same frequency. The frequency bands assigned for DAB broadcasting are different from the traditional broadcasting bands and are separated almost by a decade. The transport of the information, audio and data, also employs new concepts like audio compression (see Chapter 3). Therefore, new receiver concepts had to be developed.

Since the DAB system is fairly complex in terms of computational requirements, it was evident that there was a need to design specific, highly integrated chip-sets covering both the analogue and the digital parts of the system. These chip-sets are the building blocks for all kinds of DAB receivers and are the vital basis for cost-effective solutions. However, owing to the rapid developments in PC technology, modern PCs are able to handle the digital part of a DAB receiver.

Today, the various types of DAB receivers can be categorised as follows:

- Car radios, either audio only or audio and data receivers.
- PC-card receivers. Some of these solutions are completely hardware based, whereas other implementations decode the digital part in software on the PC.
- Home receivers including HiFi tuners and kitchen radios.
- Portable receivers.

Digital Audio Broadcasting: Principles and Applications, edited by W. Hoeg and T. Lauterbach
©2000 John Wiley & Sons, Ltd.

- Monitor receivers for network monitoring.

8.1.1 *Normative Receiver Requirements*

Several normative requirements related to DAB receivers have been standardised during the last few years. An overview of these standards is provided in Table 8.1. The scope of the various normative standards is referred to in the subsequent sections as indicated in Table 8.1.

Table 8.1 Normative receiver requirements

Title	Scope	Reference	Section
CENELEC: EN 50 248 (1997). Characteristics of DAB receivers. Brussels.	Minimum receiver requirements	(EN 50248)	8.1
ETSI: EN 300 401 3rd ed. (1999). Radio broadcasting systems: Digital Audio Broadcasting (DAB) to mobile, portable and fixed receivers. Geneva.	DAB standard	(EN 300401)	8.1
CENELEC: EN 50 255 (1997). Digital Audio Broadcasting system – Specification of the Receiver Data Interface (RDI). Brussels.	Data interface	(EN 50255)	8.5
ETSI: EN 301 234 V1.2.1 (1999). Digital Audio Broadcasting (DAB); Multimedia Transfer Protocol (MOT). Geneva.	Data services	(EN 301234)	8.5
ETSI: EN 301 700 (2000). Digital Audio Broadcasting (DAB): Service referencing from FM-RDS; Definition and use of RDS-ODA. Geneva. (still under consideration)		(EN 301700)	8.8
ETSI: TR 101 758 (2000). Digital Audio Broadcasting (DAB): DAB signal strength and receiver parameter. Targets for typical operation	Receiver requirements	(TR 101758)	8.1
ETSI: TS 101 500 (2000). Digital Audio Broadcasting (DAB): Multichannel Audio. (still under consideration)	DAB audio decoder	(TS 101500)	8.4
ETSI: TS 101 757 (2000). Digital Audio Broadcasting (DAB): Conformance Testing for DAB Audio.	DAB audio decoder	(TS 101757)	8.4

The most important among these standards are the "minimum receiver requirements", standardised by CENELEC (EN 50248), which define the minimum and typical performance of DAB receivers.

EN 50248 is the result of a joint effort of the receiver industry and experts from Eureka 147. One topic for discussion between the receiver manufacturers and broadcasters was the minimum receiver sensitivity. In a previous draft of EN 50248, a minimum sensitivity of −91 dBm for Band III and −92 dBm for L-band was assumed while the broadcasters assume for network planning receivers with −99 dBm sensitivity in an AWGN channel. The network planning (see section 7.4), as agreed upon in the Wiesbaden Conference of CEPT, follows the ITU P.370 standard (ITU-R 370, 1990) also used for TV network planning. For TV coverage, a fixed antenna at 10 m height is generally assumed. Because detailed planning data for DAB coverage were missing at the time of the Wiesbaden Conference, a correction factor of 10 dB was assumed to account for the antenna height of more or less 1.50 m which is usually used for mobile reception in vehicles.

Based on this background, the ITU still demands receivers with very good sensitivity. While the coverage in Band III is usually quite good – typical reception levels are around –70 dBm – L-band coverage is critical in many cases. This however, is mostly because of terrain effects. In a line of sight situation, or with only fairly small obstructions between the transmitter and receiver, L-band reception is sometimes possible over 50 km distance, for a planned coverage area of about 10 km radius, while in valleys or in the shadow of tall buildings reception can be impaired over a few km distance towards the transmitting site.

ITU P.370 assumes a reception level variation according to the varying receiver location of 5.5 dB which would be appropriate for an antenna height of 10 m. Because of terrain effects, the "real-world" reception level variation is usually much higher, sometimes 20 dB. Improving the receiver sensitivity by, say, 3 dB would only have a minor effect in this situation, as little as doubling the transmission power. The DAB system, however, supports single frequency networks (SFNs). By installation of several transmitters and gap fillers on the same frequency good coverage can be achieved even at L-band (see section 7.4.6).

8.1.2   Receiver Architecture Overview

Figure 8.1 presents the block schematic of a typical DAB receiver. The signal received from the antenna is processed in the radio frequency (RF) front-end, filtered and mixed to an intermediate frequency or directly to the complex baseband. The resulting signal is converted to the digital domain by corresponding analogue-to-digital converters (ADCs) and further processed in the digital front-end to generate a digital complex baseband signal. This baseband signal is further OFDM demodulated by applying an FFT (Fast Fourier Transform), see sections 2.2.1 and 8.3.2.

Each carrier is then differentially demodulated (DQPSK, see section 8.3.3) and the deinterleaving in time and frequency is performed. Finally, the signal is Viterbi decoded, exploiting the redundancy added at the transmitter side for minimising the residual error caused by transmission errors. After the Viterbi decoder, the source coded data, like audio and data services and FIC information, are available for further processing. The selected audio sub-channel is decoded by the audio decoder, whereas a data stream might be transferred to an external decoder through the receiver data interface (RDI, see 8.5) or other interfaces.

Details of each processing step are provided in the subsequent sections 8.2 (RF Front-end), 8.3 (Digital Baseband Processing), 8.4 (Audio Decoder) and 8.5 (Interfaces).

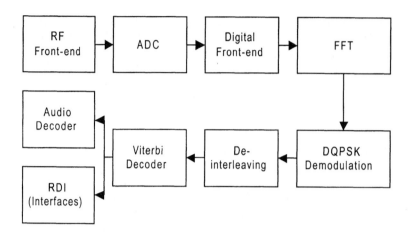

Figure 8.1 Receiver block schematic

8.2   RF Front-end

Since a COFDM-based DAB system has some special properties, traditional broadcasting receiver designs which have been designed for analogue transmission standards cannot be used. The new digital system imposes some special requirements which led to new receiver architectures.

8.2.1   Requirements

A COFDM signal can more or less be viewed as band-limited white noise. This is because the signal contains many individual carriers which are independently modulated according to the slow symbol rate, but as the individual carriers are completely uncorrelated, the signal behaves like Gaussian noise. In the frequency domain, the signal spectrum is limited according to the allocation of the sub-carriers and looks like a "Simpson's head", very similar to a WCDMA (Wide-Band CDMA) signal. In the real world, insufficient filtering and transmitter non-linearities cause "sidelobes" in the signal spectrum, limiting the adjacent channel rejection of the system (see section 7.2.6). Figure 8.2 presents the spectrum of a received DAB signal.

The special properties of COFDM impose some requirements on the receiver's RF circuits as described in the following subsections.

8.2.1.1   Non-constant Envelope
Unlike many phase-modulated systems like FM, GSM, AMPS, limiting amplifiers cannot be used, since the signal would be clipped and the amplitude part of the information would be lost. Limiting amplifiers would eliminate the requirement for gain control, greatly simplifying systems design.

A COFDM receiver requires a highly linear signal path from the antenna to the demodulator which is realised as an FFT in the digital domain, like other systems based on "non-constant envelopes", such as WCDMA.

Because of mobile operation, the average signal amplitude constantly varies by about 20 dB. Hence, for COFDM receivers, careful AGC (Automatic Gain Control) design is a vital feature.

A special scheme called a null symbol (see section 2.2.2), a break in the transmitted signal of, for example, 1 ms in DAB transmission mode I, is used for synchronisation. Special care has to be taken to pause the AGC circuit during reception of the Null-symbol.

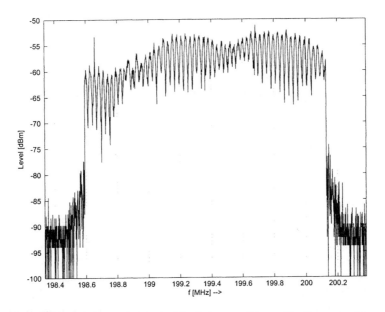

Figure 8.2   DAB signal in frequency domain. Ripple is caused by multipath propagation

8.2.1.2   Phase Accuracy

Since the Eureka 147 DAB system partly uses the same frequencies as classical analogue TV signals, conventional TV tuners might be an option for a DAB receiver for Band III.

For TV tuners, VCO (Voltage-Controlled Oscillator) phase noise is usually of no concern. The signal information is contained in the amplitude; the signal phase contains no information. Even for the signal-to-noise ratio of FM sound, phase noise is not important since in most analogue receivers the stable signal carrier of the transmitted signal is used to generate the 5 – 5.5 MHz sound carrier according to the CCITT TV standard used in many countries.

The COFDM signal consists of many individual carriers which are modulated in phase. A VCO with low phase noise (jitter) is essential for downconverting the signal from the input frequency to a proper IF. Although absolute phases are not important because of the differential phase modulation, only a total RMS phase error of less than 10 degrees can be tolerated without affecting the BER of the system. Therefore, high-performance VCO and PLL circuits are required, exceeding the specifications of those for analogue systems.

8.2.1.3 Wide Dynamic Range

DAB is a broadcasting system. An SFN may be viewed as a cellular network with each cell reusing the same frequency. Unlike cellular phone networks, where typical cell sizes are as small as a single building, a typical "cell" for Band III is 50 km in diameter, and for L-Band 15 km. Transmission powers are up to several kW, for example 10 kW for the Canadian DAB system, while base stations for cellular networks only use up to typically 10 W. This means that the receiver has to accommodate larger signals at the input, according to this difference in transmission power. While CDMA receivers are designed for a maximum input power of –25 dBm, a DAB receiver should be operational for up to –15 dBm at the input in L-band. In Band III, maximum input levels are even more critical. In this case, not just DAB transmitters with moderate power levels about 1 – 4 kW are present. Strong TV stations occupy the same frequency band, with effective transmission powers of typically 100 kW. Although the different polarisation of the signals (TV mostly horizontal, DAB vertical) provides some decoupling, it should be assumed that a nearby interferer may be up to 10 dB stronger than the strongest signal for L-band. (EN 50248) assumes a maximum input level of –5 dBm with linear circuit operation (see Table 8.2).

Table 8.2 Maximum input levels as defined in (EN 50248)

Band	Minimum	Typical
L-band	–25 dBm	–15 dBm
Band III	–15 dBm	–5 dBm

Although requirements are much harder than for mobile phone handsets, maximum signal levels are not as high as for FM reception where 2 V_{pp} at the antenna input are not uncommon in critical reception situations. Since the antenna aperture and hence the received signal level become smaller for higher frequencies, the situation is better for DAB, where the input frequency always exceeds 174 MHz. Also the TV band is usually only occupied by a single strong station, while for FM about five strong transmitters all broadcasting different programmes are often operated on a single transmitter site.

High input levels require high linearity, and high linearity is a constraint not in accord with low-power consumption or good sensitivity. A DAB receiver must be optimised more for wide dynamic range and high selectivity, while a satellite receiver (GPS) can be entirely optimised for a low noise figure only.

8.2.2 Analogue Front-end Concepts and Architectures

Besides the special signal properties of OFDM signals, the two widely separated frequency bands (see Table 8.3) bring about certain constraints for receiver design.

Table 8.3 DAB Frequency Bands

Band	Minimum	Maximum
L-band	174 MHz	239 MHz
Band III	1452 MHz	1496 MHz

The centre frequencies of both bands are separated by a factor of 7–8. Unlike TV or DVB-T tuners which can basically cover the required frequency range with the same single-conversion architecture, it is not possible to extend, for example, a UHF tuner up to 1.5 GHz.

On the other hand, the two bands are quite narrow and it is not necessary to support reception at any frequency in between. This allows the design of special "dual-band" receiver architectures.

8.2.2.1   Direct Conversion/Zero-IF

A common approach for multiband receivers is to use no IF at all and to downconvert the signal to an IF of zero (cf. Figure 8.3). By doing this, any intermediate frequency translation steps which always bring about spurious reception frequencies, and hence special filter requirements, can be avoided. This approach is very promising and is presently adopted in triple-band GSM handsets, completely eliminating the need for IF filters (Strange, 2000). Two or three switchable band select filters are used in front of the LNA (Low-Noise Amplifier). A similar architecture may be used for DAB and is presently being investigated.

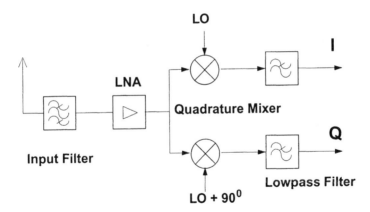

Figure 8.3 Zero-IF DAB receiver

This architecture, however, requires perfectly matched (with respect to amplitude and phase) low-pass filters in the I and Q signal path. For transmission systems with relatively small bandwidth (GSM: 20 kHz, IS95: 30 kHz), the required sample rates for the low-pass filters and ADCs are fairly low. For wide-band signals, more sophisticated low-pass filters are required. In addition, it has to be mentioned that this architecture requires two completely independent ADCs with corresponding good match.

This usually led to a design trade-off not in favour of zero-IF concepts. To the author's knowledge, no commercially available DAB receiver is based on zero-IF.

Recently, a completely integrated zero-IF baseband chip for "cdmaOne" (IS.95: 1.2 MHz bandwidth) was published (Liu, 2000), demonstrating the progress in zero-IF concepts and mixed signal CMOS integration. New developments of zero-IF-DAB receivers, for example, with RF-CMOS front-ends, can be expected, as presented for the "Bluetooth" technology.

The most severe technical problem of zero-IF receivers is LO feedthrough to the antenna input, causing a DC offset in the IQ signal. This DC component has to be removed for proper signal processing, for example by capacitive coupling. This is not feasible with every modulation scheme, but COFDM is perfectly suitable for this approach.

8.2.2.2   Receivers Based on TV Tuner Concepts

Since TV tuners for Band III are readily available, most DAB receivers basically use a modified "TV tuner" for Band III. This implies a similar IF of 38.912 MHz. The choice of the IF is determined by the required image rejection. With an IF of 38 MHz, the spurious image reception frequency for the lower end of Band III (174 MHz) would be above the upper band of the TV band, at

$$f_{image} = 174 + 2 \times f_{IF} = 250 \text{ MHz}. \tag{8.1}$$

This frequency assignment prevents strong TV/DAB stations at the upper edge of Band III acting as interferers for stations at lower frequencies. The spurious image reception frequency moves away from Band III if a frequency towards the upper end of Band III is tuned in.

Of course, if a traditional TV tuner is used for DAB reception, the IF filter has to be replaced by a proper DAB channel selection filter and the VCO has to be "cranked up" for better phase noise performance.

Support for L-band can be added by a block downconversion. The width of L-band is smaller than the width of Band III, so it is possible to downconvert the complete L-band with an LO set to a fixed frequency down to Band III.

This concept, depicted in Figure 8.4, was originally developed inside the JESSI project (Jongepier, 1996). It is used in most commercial receivers because several chip-sets based on this concept are available.

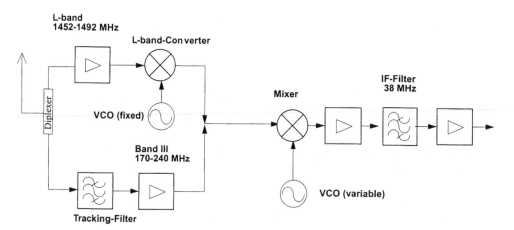

Figure 8.4   JESSI tuner

This tuner concept brings about some problems, however, especially at L-band:

1. The IF of 38.912 MHz is fairly high for direct processing by an ADC; usually a second (third in the case of L-band) IF is required
2. The first IF frequency for L-band reception (which actually is Band III) is occupied by possibly very strong interferers. If for example a −95 dBm L-band station should be tuned with a −5 dBm blasting TV station present on the same frequency, more than 90 dB isolation is required. This is technically hard to achieve.
3. Tuneable image rejection filters ("tracking filters") have to be used. These filters are bulky and are subject to parameter variations. Therefore, filters have to be manually tuned or electronically adjusted, based on stored alignment data.

Since these tuners are highly optimised for TV reception and have a long design history, performance for Band III reception is usually good.

8.2.2.3   High-IF Tuner

Some limitations of the "JESSI" tuner can be overcome by using upconversion for Band III. This concept has also recently been employed for TV tuners and set-top boxes (Taddiken, 2000). A block diagram of a tuner derived from this tuner concept is given in Figure 8.5. The high-IF technique is used for Band III. L-band is downconverted, but also to high IF.

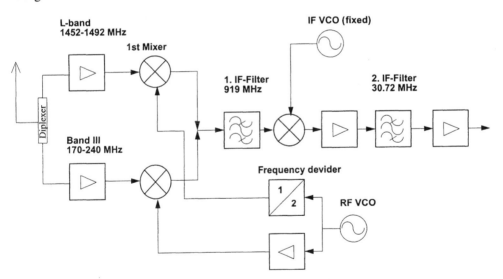

Figure 8.5 Receiver with high-IF for Band III

The most important advantage of this concept is the fact that the spurious image reception frequency for Band III is moved to very high frequencies. Therefore, no tuneable pre-selection filters are required, eliminating the requirement to manually/electronically adjust any components. With careful circuit design of the Band III LNA, tracking filters can be replaced by a band-pass filter. This enables a dramatic reduction of the form factor of the tuner, allowing compact designs for portable applications, for example PCMCIA cards.

The first IF may be chosen according to two aspects:

1.   IF should not be used by high-powered wireless services.
2.   A common VCO with a limited tuning range should be used.

A possible approach might be to use an IF at 650 MHz, exactly in the middle of both bands. This frequency is, however, also used by strong UHF TV stations, but most importantly, this arrangement would place Band III at the image frequencies for L-band reception and L-band at the image frequencies for Band III reception, again causing isolation problems. A solution is to use frequency dividers for deriving the LO frequencies from a common VCO.

In the AD6002/6003 chip-set (Titus, 1998; Goldfarb, 1998; Clawin, 2000), an IF of 919 MHz is used. This frequency allows optimum Band III image suppression, is not used by a major service (actually, it is in the gap between GSM uplink and downlink) and only requires a VCO with a small tuning range of about 10%.

Based on the concept of the AD6002/6003 chip-set, a single chip DAB tuner was presented at IFA 99 by Bosch, sized about 2 cm x 3 cm (cf. Figure 8.6). This tuner is the most compact DAB tuner design presented so far. This high-IF architecture even supports FM reception with the same receiver architecture.

Other vendors are looking at using an IF of 300 MHz, employing a 2:1 frequency divider for the Band III LO. In this case, the requirements for Band III image filters are still very stringent, since image reception frequencies are located in the crowded UHF band.

It is obvious that the number of components of such a design is already lower than for a contemporary high-performance FM tuner. It can be expected that, once high volume is reached, a DAB tuner may be manufactured for a lower cost than a state-of-the-art FM tuner.

Figure 8.6 DAB module based on single chip tuner

8.2.3   Trends, Future Developments

Since a single chip integration of the tuner is already available, the number of external components should be reduced. Integrated VCOs which have been successfully introduced in products (Taddiken, 2000) will eliminate external VCO modules. SAW filters may be either eliminated by zero-IF concepts or integrated based on "System in a Package" technologies, in combination with other external LC components. Since SAW filters provide the inherent advantage of not requiring an active power supply, they should persist in advanced receiver designs as an alternative to zero-IF.

For Bluetooth, a fully integrated single chip receiver was recently presented (Gilb, 2000). This is possible since the minimum sensitivity is only −70 dBm for Bluetooth and receiver requirements are much simpler. For good DAB signal coverage, a DAB home receiver might also work reasonably well with just −70 dBm sensitivity, but since most people are listening to radio while driving to work, such simplified receiver designs will not provide sufficient performance in the foreseeable future.

(Atkinson, 1998) gives an overview about existing technologies for receiver integration. While CMOS is making big advances, most analogue front-end chips for performance-critical applications will be BiCMOS, either with or without SiGe bipolar devices.

In the United States, COFDM-based audio broadcasting systems are presently deployed as gap fillers for future satellite-based radio systems. A technology boost can be expected if these systems find widespread acceptance.

8.3   Digital Baseband Processing

Digital baseband processing is the generic term for all signal processing steps starting directly after digitisation of the IF signal using an ADC until the source coded data become available after Viterbi decoding. In the case of DAB baseband processing includes the following processing steps:

- Generation of the complex baseband signal
- OFDM demodulation, possibly combined with compensation for the frequency drift of the baseband signal
- Demodulation of the DQPSK modulated carriers
- Time and frequency deinterleaving
- Channel decoding using the Viterbi algorithm
- Synchronisation of time, frequency and phase.

8.3.1   Digital Front-end

RF front-ends provide two different types of interfaces for the baseband processing depending on the overall receiver architecture chosen (cf. Figure 8.7a, b).

I/Q interface: In this architecture the generation of in-phase (I) and quadrature (Q) components of the complex baseband signal is done in the analogue domain. This type of interface naturally occurs in zero-IF receiver concepts, which are believed to be a path for high integration of RF front-ends for DAB (see section 8.2.2). A major disadvantage of this approach is the quality requirements for amplitude and phase balance over the required signal bandwidth of 1.536 MHz. In the context of the JESSI project these problems led to the decision to focus on the IF concept.

IF interface: Here, the last IF of the RF front-end is fed into the ADC and frequency correction is provided via a complex multiplier fed by a numerical controlled oscillator (NCO). After shifting the signal towards zero frequency low-pass filters are employed to provide image and adjacent channel suppression. Finally, the signal is decimated towards the sampling rate of F_c = 2.048 MHz. Owing to the digital implementation of this architecture, phase and amplitude imbalance requirements can easily be solved. In the following we focus on the IF interface.

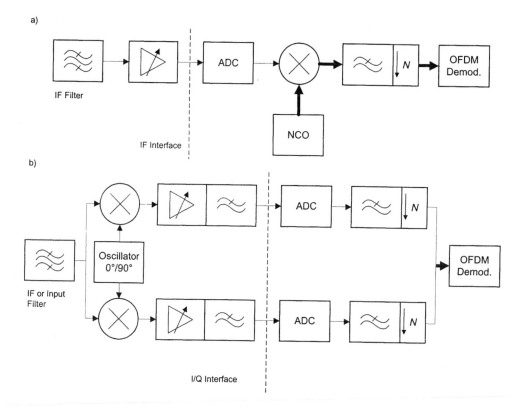

Figure 8.7 Receiver architectures and interfaces: a) intermediate frequency interface; b) in-phase (I)/ quadrature (Q) interface

In order to minimise the hardware implementation cost, the sampling rate of the ADC is usually chosen as an integer multiple, N, of the sampling frequency of the complex baseband signal F_c = 2.048 MHz, that is

$$F_{ADC} = N F_c. \tag{8.2}$$

The actual choice of the ADC sampling frequency is a trade-off between filtering in the analogue and the digital domain. The minimum possible ADC sampling frequency for an IF concept occurs for $N = 2$ with $F_{ADC} = 4.096$ MHz, a choice which can be found in the early JESSI concepts. An advantage of the IF concept in Figure 8.7a is the fact that the IF can freely be chosen according to the needs of the special front-end architecture. Any IF can be supported as long as it is compatible with the ADC chosen regarding input bandwidth and sampling rate. However, more attractive solutions – regarding hardware implementation costs – can be found if certain relations between ADC sampling rate and IF are fulfilled. This will be shown in section 8.6.2.2.

8.3.2   OFDM Demodulation

The demodulation of the OFDM symbols is performed by applying FFTs to calculate the complex amplitudes of the carriers of the DAB spectrum. These amplitudes contain the information of the modulated data by means of a DQPSK modulation. A complete overview of OFDM demodulation including the synchronisation function is given in Figure 8.8. According to the various DAB transmission modes I–IV (see section 2.2.1), FFT lengths varying from 256, 512, 1024 and 2048 have to be implemented as indicated in Table 8.4. This can be realised very efficiently by the well-known Radix-2 FFT algorithm using a simple control of the FFT memory addressing.

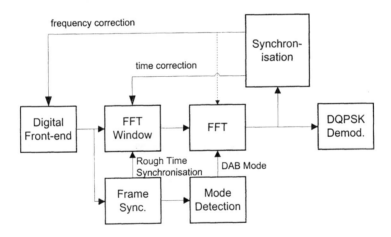

Figure 8.8   OFDM demodulation

Table 8.4   DAB transmission modes and FFT length

Mode	Carriers	FFT Length
I	1536	2048
II	384	512
III	192	256
IV	768	1024

To cope with the possible frequency drift of the baseband signal, an AFC (Automatic Frequency Control) is necessary. One possible realisation is indicated in Figure 8.7a, where the frequency shift is compensated by means of a complex mixer stage and an NCO. An even more attractive solution, which avoids the complex mixer stage, takes advantage of a modified FFT algorithm and is described in section 8.6.2.2.

8.3.3   DQPSK Demodulation

Differential demodulation of the used carriers is usually performed by applying a complex multiplication with the stored complex conjugated amplitude of the last OFDM symbol. Initialisation of this process is done using the phase reference symbol (TFPR) (see section 2.2.2). Figure 8.9a gives an overview of the algorithmic processing steps and indicates typical word widths encountered in hardware implementations. Figure 8.9b gives an example of a possible mapping of the demodulated amplitudes to soft decision values suitable as inputs to the Viterbi decoding algorithm, cf. section 8.3.5. The parameter σ is used to adapt the characteristic curve to the actual signal level in the receiver chain.

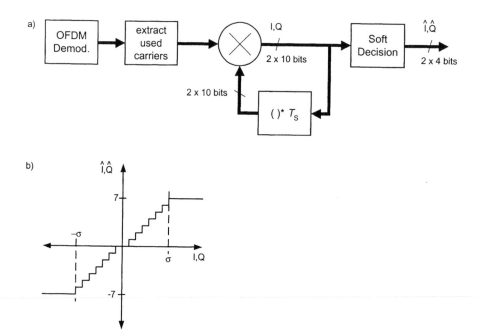

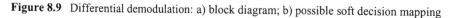

Figure 8.9 Differential demodulation: a) block diagram; b) possible soft decision mapping

8.3.4   Deinterleaving

In order to cope with transmission disturbances, two interleaving mechanisms are used (see section 2.2.4):

- Frequency interleaving is a rearrangement of the digital bit-stream over the carriers, eliminating the effects of selective fades. Frequency interleaving operates on one OFDM symbol only.
- Time interleaving is used to distribute long error bursts in order to increase the channel decoder's error-correcting capability.

The frequency deinterleaving can be implemented by addressing the output of the FFT according to the interleaver tables.

The time deinterleaving is a task that requires a substantial amount of memory. As presented in Chapter 2, the data of each sub-channel are spread over 16 CIFs, whereas each CIF represents the information of 24 ms. Thus the interleaving process requires a memory that has 16 times the capacity of the data to be decoded.

As an example, we examine an audio sub-channel with the typical bit rate of 192 kbit/s. One audio frame of 24 ms duration equals 576 bytes. Since the time deinterleaving is located prior to the Viterbi decoder, each information bit is represented by its soft decision value, typically a 4-bit number. Thus, the memory required for deinterleaving this sub-channel works out to 36864 bytes.

The maximum amount of memory needed for time interleaving and assuming the storage of 4-bit soft decision output values of DQPSK demodulation works out to 442 kbytes or 3.54 Mbits. This amount can be halved by using appropriate in place usage of this memory leading to a necessary amount of 221 kbytes or 1.77 Mbits for a full-stream DAB decoder.

8.3.5   Viterbi Decoding

To combat errors due to channel distortions, DAB employs a powerful punctured convolutional code (RCPC) with constraint length 7 and mother code of rate ¼ for channel coding. This mother code is punctured (see section 2.2.3) to obtain a wide range of possible code rates so as to adapt the importance of the information bits to the channel characteristics. For decoding these codes, the Viterbi algorithm is used (Proakis, 1995), which offers the best performance according to the maximum likelihood criteria.

The input to the Viterbi decoder can be hard-decided bits, that is '0' or '1', which is referred to as hard decision. A better performance (2.6 dB improvement) is achieved if the uncertainty of the input is known to the Viterbi decoder, by using intermediate values. The optimum performance for this soft decision is reached when each input value is represented by a 16-bit number. However, the degradation is still negligible if the number of bits is reduced to 4 bits (Proakis, 1995).

The energy dispersal de-scrambling is another task that can easily be assigned to the Viterbi decoder module. The BER (Bit Error Rate) on the channel can be estimated by re-encoding the decoded sequence or a sub-set of the sequence and comparing this sequence with the received bit-stream (see section 7.5.2). This information can be used as additional reliability information.

8.3.6   Synchronisation

Synchronisation of a DAB receiver is performed in several steps:
1. Coarse time or frame synchronisation
2. Coarse frequency synchronisation on carrier accuracy
3. Fine frequency synchronisation on sub-carrier accuracy
4. Fine time synchronisation.

Frame synchronisation. The Null-symbol of the DAB transmission frame provides a simple and robust way for the coarse time synchronisation, which is also called frame synchronisation. The underlying idea is to use a symbol with reduced signal level which can be detected by very simple means. In practice a short time power estimation is calculated which is then used as input to a matched filter. This filter is simply a rectangular window with a duration according to the Null-symbol length. Finally a threshold detector indicates the beginning of a DAB frame. It is also possible to calculate an AGC value for optimal scaling inside the following FFT signal path (FFT stages).

Coarse and fine frequency synchronisation. Coarse and fine frequency synchronisation can be performed using the TFPR symbol in the frequency domain. This step clearly requires a sufficiently exact coarse time synchronisation. Frequency offsets are calculated using the various CAZAC (Constant Amplitude Zero Autocorrelation) sequences inside the TFPR symbol. These sequences provide a pulling range of the AFC of about ±32 carriers. This is a sufficiently large value to cope with cheap reference oscillators used in RF front-ends.

Fine time synchronisation. Fine time synchronisation is performed by calculating the channel impulse response based on the actually received TFPR symbol and the specified TFPR symbol stored in the receiver.

All the described steps are subject to algorithmic improvements and contain various parameters which reflect the receiver manufacturer's experience in the field. Thus in all concepts which are on the market today synchronisation is mostly performed in software on a digital signal processor (DSP).

8.4   Audio Decoder

As presented in Chapter 3, the audio coding scheme used in DAB is MPEG-1 and MPEG-2 Layer II (IS 11172, IS 13818). For DAB use, these standards have been extended to provide further information for the detection of transmission errors in those parts of the bit-stream with the highest error sensitivity. This is useful for error concealment. Furthermore, the system provides a mechanism to reduce the dynamic range of the decoded audio signal at the receiver which is useful, especially in noisy environments like vehicles.

8.4.1   Audio Decoder Architecture

The DAB audio decoder is based on an MPEG-1 and MPEG-2 Layer II decoder, but can additionally calculate and utilise error status information of the audio bit-stream like ISO-CRC and SCF-CRC (see Chapter 3) which is necessary for a powerful error concealment.

Furthermore, the decoder should be able to decode dynamic range control (DRC) information. However, the DAB specific extensions to the MPEG audio standard are not normative and the receiver manufacturer can even decide not to exploit this information. The block schematic of a DAB audio decoder is depicted in Figure 8.10.

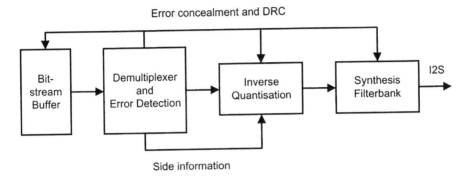

Figure 8.10   Audio decoder block schematic.

As presented in Chapter 3, an audio frame consists of header and side information data, sub-band samples, error detection information and additional data that are intimately related to the audio programme, like PAD and X-PAD.

Firstly, the header is decoded and information like bit rate and audio mode are extracted. Based on this information, the table is selected that is necessary to interpret the side information correctly. The side information contains the quantisation that was employed in the encoder and the scaling factor that was used for normalisation for each sub-band. In MPEG Layer II audio coding, the audio signal is transformed into the frequency domain and is decomposed into 32 equally spaced sub-bands. The inverse process is the synthesis filterbank. The reconstructed 32 sub-band samples in the frequency domain are transformed into 32 consecutive samples of one channel in the time domain, applying a 32-point IMDCT followed by a windowing process. Finally, the audio signal is transferred to a digital-to-analogue converter (DAC) using the I2S protocol (see section 8.5).

The synthesis filter is the most demanding task with respect to the computational effort. Consequently, there exist a number of algorithms that offer more efficient solutions in terms of multiply and add operations, but the drawback is that the addressing becomes more complex.

8.4.2   *Normative Requirements*

Conformance testing for DAB Audio (TS 101 757) defines the normative requirements for a DAB audio decoder which is based on the procedures defined by MPEG as part 4 of (IS 11172) and (IS 13818) respectively. (TS 101 757) defines a test procedure and associated bit-streams that can be used to verify whether an audio decoder meets the requirements defined in (EN 300 401). The basic principle of the test is depicted in Figure 8.11.

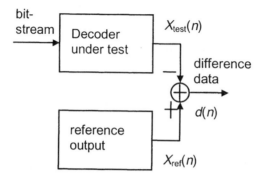

bit-
stream | Decoder under test | $X_{test}(n)$

difference
data

$d(n)$

reference
output

$X_{ref}(n)$

Figure 8.11 Audio conformance test procedure

For the calculation of the conformance criteria, the definition of the difference of the two related output samples $(d(n))$ and their error energy (RMS) are given below

$$d(n) = x_{ref}(n) - x_{test}(n) \tag{8.3}$$

$$RMS = \sqrt{\frac{1}{N} * \sum_{n=0}^{N-1} d(n)^2} \tag{8.4}$$

The decoder under test decodes the test bit-stream. The output (x_{test}) is digitally recorded and compared on a sample-by-sample basis with the reference output (x_{ref}). Two certificates for DAB audio decoder are defined:

1. "full accuracy DAB audio decoder", which is more strict
2. "DAB audio decoder", which is more relaxed.

To be called a "full accuracy DAB audio decoder", the average error energy as defined in equation (8.4) must be below $2^{-15/\sqrt{12}}$ and in addition the first 14 bits of each output sample must be bit exact. If the average error energy does not exceed $2^{-11/\sqrt{12}}$, the criteria for a "limited accuracy DAB audio decoder" are fulfilled.

8.5   Interfaces

In principle, DAB receivers can be equipped with two types of interfaces: those that carry data to or from DAB receivers and those that carry control information.

Data interfaces. The RDI (Receiver Data Interface) (EN 50255) is a data interface which was specifically developed for DAB. It is suitable for connecting an external data decoder to a DAB receiver. It carries the full DAB stream of up to 1.8 Mbit/s and is thus useful for high data-rate applications, like video broadcast through the DAB system. Nowadays, this

interface is less important, since a PC-card receiver transfers data through the PC bus system, like PCI, and for car radios there are other specific solutions, like MOST.

Control interfaces. The I2C (Inter IC) interface belongs to the second category. It is used to exchange control information, like the frequency, that the front-end should tune in. The bit rate of this I2C is up to 400 kbit/s. In this context, this interface is internal to the DAB receiver and connects the digital part or a microcontroller with the analogue front-end.

Control interfaces are also used to control a black-box-type DAB receiver (see section 8.7) with the car radio. APIs to control a DAB receiver box through car bus systems, like MOST, are currently defined.

Table 8.5 provides an overview of the most important interfaces of DAB components and receivers.

Table 8.5 Overview of interfaces to DAB receivers

Interface	Area	Description
RDI (Receiver Data Interface)	Data and audio	This interface is defined in (EN 50255). It is the standard interface of DAB receivers that can transmit the content of the complete DAB ensemble of up to 1.8 Mbit/s.
I2S (Inter IC Sound)	Audio	Audio interface for digital audio data in the time domain (PCM format). It is used typically as the interface to the ADC.
AES/EBU S-P/DIF	Audio	Audio interface for PCM coded data. AES/EBU is the professional format, whereas S-P/DIF is the consumer format. Physically, both protocols are based on (IEC 60958). The difference is in the interpretation of some of the "side information bits". For further details we refer the reader to (IEC 60958) and (Watkinson, 2000).
IEC 61937	Audio	This interface allows compressed audio to be carried over the an interface, that is physically identical to (ISO/IEC 60958). This allows for example a multichannel audio programme received over DAB to be feed to an external multichannel decoder. For further details we refer the reader to (IEC 61937).
I2C (Inter IC)	Front-end	This is a typical three wire interface to control the analogue front-end of the DAB receiver. The upper limit for the data rate is about 400 kbit/s.
SPI	Front-end/data	An alternative to I2C allowing higher data rates. It can in addition be used to transfer any data (like data services) to any external device.
MOST	General	In-car bus system. One application is to link the DAB receiver to the front unit that controls the receiver.
IEEE 1394 (FireWire)	General	Alternative bus system to MOST that allows higher data rates. IEEE 1394 is even suitable for uncompressed video distribution.

8.6   Integrated Circuits for DAB

Chip-sets are the building blocks for the various kinds of DAB receivers. From the very beginning of DAB standardisation in 1986 it was clear that the delivery of key components for receivers, that is highly integrated circuits – ICs, is substantial for the success of the new standard. Today the integration is at a level where a complete DAB and FM broadcast receiver can be realised using two highly integrated circuits (RF chip and baseband chip). It is expected that in the near future (2–4 years) single chip solutions using SiGe technology will enter the market and allow the production of lowest cost DAB receivers.

8.6.1   *Overview of Existing DAB Chip-sets - the JESSI Programme*

The European JESSI (Joint European Submicron Silicon Initiative) programme was a suitable umbrella for the necessary pre-competitive activities of the European semiconductor and equipment industry. Inside JESSI the AE-14 programme was one of the application-specific programmes and was started in 1991 at a time when the final DAB specification, now officially known as (EN 300401), was not available. The system specification and verification, however, had already been done during the Eureka 147 project. Most of the partners which were active in the Eureka 147 project also played an important role during the JESSI AE-14 project. The overlap of the two groups ensured that AE-14 had early access to the DAB system specification and an extremely thorough understanding of the DAB system.

    The JESSI AE-89 DDB (Digital Data Broadcasting) project was launched in 1995 in order to support the introduction of data services into DAB by delivering an IC at a very early stage. In addition to this, the combination of DAB and FM and the further integration of DAB have been focused by this project.

    The following paragraphs give an overview of the final silicon delivered by AE-89 which was the basis of numerous receiver designs in the past few years.

8.6.1.1   Analogue IC – JESSI AE-89

As mentioned above, most DAB tuners are based on the "JESSI concept" as shown in Figure 8.4. In the first implementation, four ICs were required. The signal strips for L-band and Band III were each integrated in a special IC, and both the L-band downconverter and the Band III signal strip required a separate PLL IC which had to be compliant with the Eureka 147 system requirements. In general, an external low noise front-end transistor was required for both bands. The popular TEMIC chip-set presented in Table 8.6 relies on this concept.

Table 8.6 JESSI analogue IC chip-set, by TEMIC

U2730 B	L-band downconverter
U2750 B	Band-III-tuner

    There are improved versions of this chip-set available, the U2730BN and U2731B. These chips incorporate the PLL circuits and active VCO circuits required, so only two ICs are necessary for a DAB tuner design. Both ICs require an external LNA transistor in order to achieve optimum performance. Controlling the tracking filters for Band III is a nontrivial task; the U2731B contains all necessary DACs to automatically generate the control voltages required (U2731B).

8.6.1.2   Digital IC – JESSI AE-89

The final generation of AE-89 DAB ICs achieved the following technical features:

- Decoding capabilities for the full DAB bit-stream (up to 1.7 Mbit/s)
- Digital I/Q generation and filter for adjacent channel selectivity
- Digital AFC
- Combined analogue and digital AGC
- RDI capable of transporting up to 1.7 Mbit/s.

For the realisation of the DAB channel decoder two architectures have been produced by TEMIC and Philips. These architectures are the result of trading flexibility against level of integration. Architecture I, developed by Philips, is shown in Figure 8.12 and offers a high level of integration.

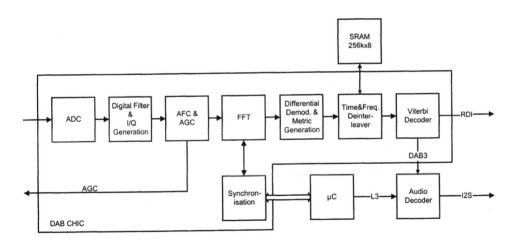

Figure 8.12   JESSI AE-89 DAB Channel Decoder Architecture I - Philips

The synchronisation functions are completely integrated and realised mainly in hardware. Closing of the synchronisation loops, however, is under the control of an external microcontroller (µC). As external components only a low-cost µC and 2-Mbit SRAM are needed.

Architecture II, developed by TEMIC and depicted in Figure 8.13, offers a lower level of integration, but has the advantage of offering more flexibility for realising different synchronisation strategies, due to the fact that the synchronisation is done in software on an external DSP.

These chip-sets have been the basis for numerous receiver designs and have been available as prototypes since the end of 1996.

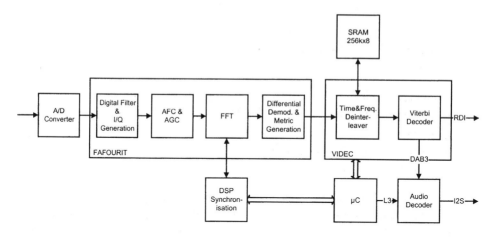

Figure 8.13 JESSI AE-89 DAB Channel Decoder Architecture II - TEMIC

8.6.2   D-FIRE Chip-set (Bosch)

The following paragraphs give a detailed description of the Bosch D-FIRE chip-set (Bolle, 1998; Clawin, 1998; Clawin, 2000), which differs in many aspects from the chip-sets which are based on the JESSI concept (see sections 8.6.1.1 and 8.6.1.2). This concept had some properties which hindered a low-cost, low-power and small-size DAB receiver. These properties are:

- The analogue VCXO required for the synchronisation is a cost-intensive component, which furthermore requires a DAC. In addition, it is well known that the VCXO control signal itself is vulnerable to interference.
- The use of tracking filters for interferer and image suppression in Band III is a major cost-producing factor: bulky, high-Q, tuneable inductors have to be used in connection with varactor diodes, which have to be driven by a control voltage generated by DACs.
- Manual or automated tuning has to be employed during manufacture of the DAB receiver.
- Supply voltages in excess of 5 V have to be used.
- A large number of VCOs are required, for example four VCOs for a receiver covering Band III and L-band.
- The L-band is sensitive to interferers at frequencies up to 300 MHz, which are ubiquitous from digital equipment (on-board DSP, PCs, TV stations).

Derived from this the following major goals for the development of this DAB receiver engine (D-FIRE: DAB–Fully Integrated Receiver Engine) chip-set were identified as:

- Use of supply voltages < 5 V for the analogue part and 3.3 V for the digital part.
- Design of an adjustment-free receiver.
- Full digital synchronisation (time, frequency and phase).

- Minimisation of the total number of VCOs.
- Low power consumption in comparison to state-of-the-art receiver concepts.

8.6.2.1 Front-end Chip-set

The commonly used receiver for DAB signals employs a Band III (174–239 MHz) heterodyne architecture, preceded by a 1.5 GHz to 210 MHz block converter for L-band (1452–1492 MHz) reception. This architecture requires tuneable pre-selection filters to meet the stringent selectivity and dynamic range requirements (EN 50248).

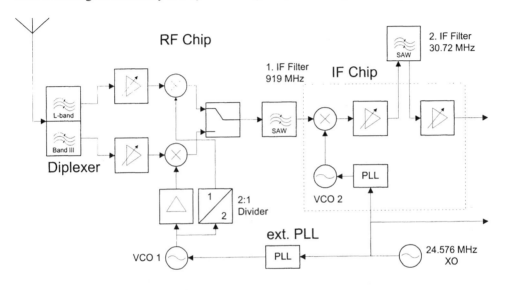

Figure 8.14 Front-end architecture of the D-FIRE chip-set.

The alternative architecture used in the D-FIRE concept exploits a high level of integration by using two separate receiver stages for the two bands, with one PLL operating both mixers to the common first IF, on a single RF chip (cf. Figure 8.14). The common IF is chosen between the two receiving bands (Band III and L-band). This choice allows the elimination of all tracking filters in Band III due to the underlying upconversion principle.

This separation allows independent optimisation of receiver designs for noise and intermodulation (IM) performance. A second PLL then performs the downconversion to the second IF of 30.72 MHz which is finally sub-sampled by the ADC using a sampling frequency of 24.576 MHz.

8.6.2.2 Digital IC (D-FIRE)

The architecture of the digital IC (D-FIRE) consists of two main parts (see block diagram in Figure 8.15):

- The signal path, which provides all the DAB functions for signal processing and decoding.

- A control block consisting of an OAK DSP core, an MIPS RISC core, 42 kbytes fast
  internal RAM for the processors and a cross-connector which realises flexible links
  between all components.

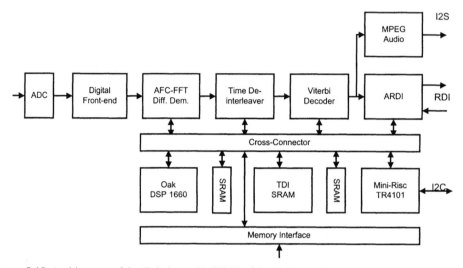

Figure 8.15 Architecture of the digital part (D-FIRE) of the DAB receiver system

The digital front-end. In contrast to the generic front-end architecture of Figure 8.7a, the
D-FIRE digital front-end employs a close relation between ADC sampling rate and IF in
order to drastically reduce the hardware implementation effort. The basic idea is to use a
complex recursive band-pass filter and subsequent decimation to generate the complex
baseband signal and to provide simultaneously adjacent channel suppression. Very little
hardware implementation effort is necessary if the following relation holds:

$$F_{IF} = \frac{F_{ADC}}{4} \cdot (1 + 2\,m) \quad \text{for } m = 1,2,\dots \ . \tag{8.5}$$

For D-FIRE the choice $m = 2$ leads to an F_{IF} of 30.72 MHz which is sub-sampled by the
ADC working at a sampling rate of $F_{ADC} = 24.576$ MHz, cf. Figure 8.16. Besides the chosen
IF the front-end is of course able to cope with IFs of 6.144 MHz, 18.432 MHz and 43.008
MHz respectively. Care, however, has to be taken in the RF front-end to provide sufficient
filtering at these unintentional IFs.

Because of the differential modulation scheme chosen in DAB it is possible to employ
recursive filters (Infinite Impulse Response – IIR) in order to meet the demanding adjacent
channel suppression requirements imposed by (EN 50248). In order to cope with the
stability requirements encountered in recursive filtering the well-known wave digital filter
(WDF) concept (Fettweis, 1986) has been employed. In particular, for the first filter a
complex polyphase filter has been chosen, which can be very efficiently combined with the
subsequent decimation by a factor of 12.

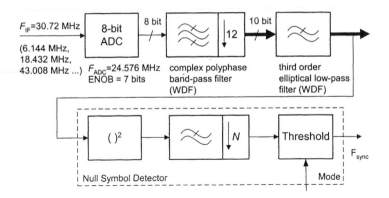

Figure 8.16 D-FIRE digital front-end

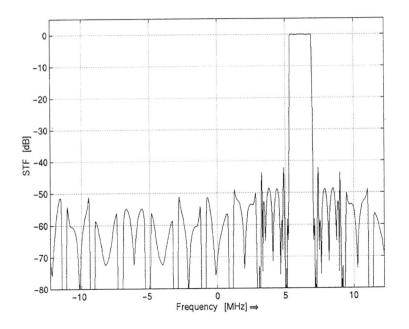

Figure 8.17 Effective transfer function of the D-FIRE digital front-end

After decimation, an additional low-pass filter is necessary to suppress the remaining interferences. For this filter a third-order elliptical low-pass WDF has been designed. Figure 8.17 shows the effective signal transfer function of the digital front-end. The design goal of this filter architecture has been to provide a minimum attenuation to possible adjacent DAB channels of at least 48 dB. It can easily be observed in Figure 8.17 that this goal has been achieved. The maximum signal-to-noise ratio of this architecture is approximately 43 dB which compares nicely to the theoretical value obtained by an ideal 7-bit ADC (35 dB SNR) and the gain obtained from the – complex – oversampling ratio of 6 (7.8 dB).

OFDM demodulation. The demodulation of the OFDM symbols is performed by applying FFTs to calculate the complex carriers of the DAB spectrum (see Figure 8.17). These carriers contain the information of the modulated data. A Radix-2 FFT with decimation in time is implemented so that the specified DAB transmission modes can be realised by a simple control of the FFT addressing. To cope with the frequency drift of the baseband signal, an AFC (Automatic Frequency Control) is necessary for which a new approach has been chosen. Let $x(k)$ be the complex output signal of the digital front-end, N the FFT length and ρ be the normalised frequency deviation, that is the measured frequency deviation normalised to the sub-carrier distance. In this case the following frequency correction has to provide the corrected signal:

$$y(k) = x(k)\, e^{-j\, 2\pi\, k\rho\, /N} \tag{8.6}$$

The Fourier transform Y of y is given by

$$Y(l) = \sum_{k=0}^{N-1} y(k)\, e^{-j\, 2\pi\, kl\, /N} \;.\; = \sum_{k=0}^{N-1} x(k)\, e^{-j\, 2\pi\, k(l+\rho)\, /N} \tag{8.7}$$

It is important to note that $Y(l)$ is now calculated by a modified discrete Fourier transform, where the running index l is replaced by $\rho + l$. The interesting point is that now fast algorithms can be derived for this modification which leads to a hardware architecture which is a slight modification of the well-known Radix 2 decimation in time FFT architecture (Bolle, 1997).

The baseband energy, calculated by the Null-symbol detector (cf. Figure 8.17), is used for optimal scaling of the fixed-point FFT implementation and the following signal path.

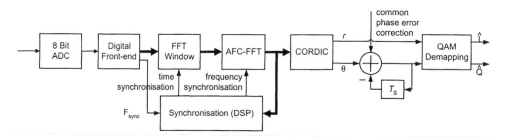

Figure 8.18 D-FIRE OFDM demodulation

Demodulation. In contrast to the approach taken by JESSI (cf. section 8.6.1.2) the DQPSK demodulation is performed by calculating the polar representation $(r,\ \theta)$ of the carriers using the CORDIC algorithm, which can be implemented very effectively in hardware compared to the costly implementation of the complex multiplier which is needed in the traditional approach. In doing so, the differential phases of two DAB symbols can easily be calculated by a subtraction of the carrier phases. In addition the phase representation allows additional functionalities such as the removal of a constant phase error which is introduced by the phase noise contribution of the RF front-end oscillators. To determine the metrics for the following soft-decision channel Viterbi decoder, the actual amplitude and the difference between two following phases is used.

Deinterleaving. The time deinterleaver is implemented in combination with a sub-channel demultiplexer. Only DAB capacity units with a size of 64 bits, belonging to selected sub-channels, are deinterleaved and transferred to the Viterbi decoder. This procedure permits a multiplex reconfiguration to be followed for all selected sub-channels in the ensemble simultaneously without any interruption or disturbance.

Viterbi decoding. To combat errors due to channel distortions, DAB employs a convolutional code with constraint length 7 for channel coding. It can be punctured to obtain a wide range of possible code rates so as to adapt the importance of the information bits to the channel characteristics. Decoding is done using the Viterbi algorithm (traceback version). The Viterbi decoder is able to decode the full data rate of 1.8 Mbit/s which is the maximum data rate specified by ETSI. All specified code rates between 8/9 and 1/4 are supported, that is all equal error protection (EEP) and unequal error protection (UEP) profiles can be decoded. The decoder operates with 4-bit soft-decision metrics and closely approaches the theoretical coding gain. Furthermore, the Viterbi module includes the calculation of the CRC for the FIC data, an energy dispersal descrambler and a re-encoding unit for estimating the BER on the channel.

Audio decoder. The architecture of the implemented audio decoder is depicted in Figure 8.19. Three major blocks can be identified: demultiplexing, reconstruction of (frequency domain) sub-band samples and synthesis filtering. The latter blocks share an MAC (Multiply Accumulate) unit that allows one MAC operation per cycle.

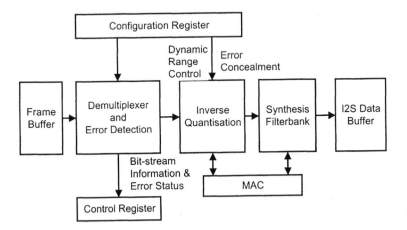

Figure 8.19 Audio decoder block schematic

A RAM is located before the decoder that stores the MPEG coded audio data. This buffer handles one part of the error concealment strategy (see section 3.7.2): in the case when the audio frame is not decodable owing to transmission errors the audio decoder requests another audio frame so the previously decoded frame is repeated. The communication between buffer and audio decoder is realised by an intelligent and flexible interface. With this buffer concept it is possible to decode ancillary data information carried in the DAB specific data field in the audio frame. Flexible concealment is guaranteed since the sub-band reconstruction process can be influenced on a sub-band basis, that is the audio signal can be shaped in the frequency domain prior to transformation into the time domain. Using this mechanism, the audio decoder can be configured in such a way that very annoying 'birdies' caused by an error-prone environment are avoided while the reproduction of the audio signal is maintained.

The audio decoder is able to use directly the information for the reduction of dynamic range which is part of the coded audio data. These data are applied during the decoding process in the frequency domain so the effect of the DRC process is smoothed by the filterbank. The synthesis filter is the most demanding task with respect to computational effort. The reconstructed sub-band samples in the frequency domain are transformed into the time domain by applying a 32-point IMDCT (Inverse Modified Discrete Cosine Ttransformation) followed by a windowing process.

8.7   Receiver Overview

The first commercially available receivers were presented at the Internationale Funkausstellung (IFA) in Berlin in 1995. These so-called fourth-generation receivers consist of a car radio that provides a tuner for the traditional analogue broadcast systems (FM and AM) and the control interface to a black box, which facilitates the complete DAB specific processing. An antenna capable of coping with both DAB frequency bands completed these receivers. They were the first DAB receivers to decode an arbitrary multiplex configuration and some were even able to follow a multiplex reconfiguration. These first car receivers, manufactured by Bosch/Blaupunkt, Grundig and later by Pioneer, were used for DAB pilot projects in Germany and other European countries.

At the IFA in 1997 about 10 manufacturers presented DAB car radios and even the first tuners fitting a 1-DIN slot have been displayed.

At the IFA in 1999 and on other occasions, the first HiFi tuners and PC-card-based receivers were presented. Today, DAB receivers for a wide range of markets are available including car radios, HiFi tuners, PC receivers, portables and professional devices. The following tables provide a snapshot of the situation in mid 2000. An up-to-date overview of DAB products can be found at (www.worlddab).

8.7.1   Car Radios

Car radios can be categorised by their capabilities; that is, whether these are audio only or audio and data receivers. Furthermore, there are so-called 1-DIN tuners where the DAB receiver is integrated in the car radio, whereas the black box concept consists of a separate DAB receiver box that is controlled by a front unit (car radio). Table 8.7 provides an overview of the first DAB car receivers that were available as prototypes or on the market.

Table 8.7 DAB car receivers

Manufacturer	Product	Characteristic
Bosch/ Blaupunkt	Hannover DAB 106 D (1995)	including data terminal screen
	D-Fire1 box (1998)	black box
Clarion	DAB 9475R, 1998	1-DIN tuner
	DAH 9500z, 1999	black box
Delphi/Delco	DAB 100, 1997 (Prototype)	1-DIN tuner
Grundig	DAB-T-1002, 1995	including data terminal screen
	DCR 200, 1998	black box
JVC	KT-DB1500	black box
Kenwood	KTC-959DAB, 1999	black box
Oritron	DAB-021681, 1999 (Prototype)	1-DIN tuner
Panasonic	Panasonic	black box
Pioneer	GEX-P900DAB, 1998	black box
	GEX-P900DAB-II, 1999	black box
Sony	XT-100DAB	black box
Technisat	CARDAB1, 1999 (Prototype)	1-DIN tuner
VDO Dayton	MS-4000, 1999 (Prototype)	navigation with DAB option

8.7.2   HiFi Tuners

The first home receivers, most of them integrated in HiFi tuners, were presented in 1998. An overview is given in Table 8.8.

Table 8.8 DAB home receivers

Manufacturer	Product	Characteristic
Arcam	Alpha 10 DRT, 1998	
Cymbol	C-DAB I	
Grundig	Fine Art Audio Tuner (Prototype)	external DAB module
Sony	Sony-ST- D777ES, 2000	DAB and FM tuner
TAG McLaren	T32R DAB Tuner, 1999	
Technics	ST-GT 1000 DAB Tuner	DAB/FM/AM tuner
Technisat	DAB TechniDAB1 (Prototype)	DAB only

8.7.3   PC-based Receivers

PC-card-based receivers offer the full variety of multimedia services that DAB provides. The first of these receivers, were presented in 1998. An overview is given in Table 8.9.

Table 8.9   DAB PC-card receivers

Manufacturer	Product	Bus	Characteristic
Bosch/ Blaupunkt	DAB-core PCI	PCI	supports data services
Psion	Wavefinder	PCI	
Radioscape	RS-T1000	PCI	monitor receiver
TechnoTrend	1998	PCI	
Terratec	Dab2PCI, 2000 (Prototype)	PCI	

8.7.4   Other Receivers

Portable receiver. The first prototype of a portable DAB receiver was presented in 1998 by Bosch. Nowadays, because of the availability of the highest integrated receiver modules, more portable DAB receivers are offered.

Reference receiver. Reference receivers are an indispensable necessity for network monitoring and coverage evaluation (see section 7.5). Information such as signal strength, BER or channel impulse response is displayed by such receivers.

8.8   Operating a DAB Receiver - the Human–Machine Interface

The main focus when designing the user interface to the receiver is to guarantee user-friendly operation. Since DAB is a new broadcast system that behaves differently to the traditional FM broadcast system, this requirement is even more important. DAB also allows for data broadcasting and this enhances tremendously its capabilities compared to the analogue broadcasting systems. This is completely new in this application area. New techniques to control and use these new capabilities had to be developed.

This Chapter concentrates on car receivers, since this is the most important, and with respect to the user interface, also the most demanding application area for digital radio. The most important point is the safety of the car driver and everything else must be subordinated to achieve this goal. This means in particular that visual control of the receiver must be minimised as much as possible and that the use of a voice recognition system should be considered.

Some of the research work has been carried out in a joint effort between receiver manufacturers, broadcasters and their research centres under the umbrella of the HuMIDAB (Human–Machine Interface DAB) project. This project was funded by the European Commission. One outcome of this project are the requirements for a user-friendly and easy-to-use DAB receiver:

- The HMI must look simple.
- Displays should provide explicit confirmation feedback.
- A "tell-me-more" function is helpful. This finally led to the development of an "electronic user manual", which provides further information on the selected feature. This additional information is context sensitive, that is dependent on the situation, and can be text based or even spoken.
- DAB should enhance existing radio rewards rather than turn into a completely new medium. The first DAB receivers mirrored the structure of the DAB system in the user interface. The difficulty is that the user first has to understand that the DAB system typically offers more than one audio service at one frequency. Some of the very first users of the DAB system complained that they received only one or two audio services. The reason was that they used the automatic scan, which automatically plays the first audio service of a DAB ensemble. To switch to another audio service of the same DAB ensemble, a new control command is necessary; this does not feature the traditional analogue broadcasting systems.
- Text and especially scrolling text should be used very carefully.
- Acronyms and icons may confuse the user when their meaning is not clear.
- For complex menu structures, an undo or home button is necessary.

In the following, we will present some basic concepts which are used in all HMIs.

8.8.1   Programme Type (PTy)

The assignment of one or more programme types (PTys) by the service provider allows selection of a programme, matching the interests of the user (see also section 2.5.4.2). PTy is a feature that is known from RDS, but for DAB use the capabilities have been largely extended. DAB offers the possibility to assign more than one PTy to a programme and to further refine a coarse PTy by a wide choice of fine PTys. For example, the coarse code "Sport" can be refined by a fine code "Football". Another novelty is dynamic PTy codes that can be used to describe, for example, the song that is currently on-air, whereas the static PTy reflects the flavour of the service itself. Therefore, the HMI has in principle to offer three modes:

Service search. Search for services with a specific flavour. That is, the static PTy of the service matches the user's choice.

Programme search. Search for services with a specific programme. That is, the dynamic PTy of the service matches the user's choice.

Background scan (watch mode). In principle, one of the above, when there is no service available that offers the programme, according to the user's choice. The programmes are watched and if a programme according to the scan-list is turned on, the receiver automatically switches to that programme.

8.8.2   Announcements

Announcements (see also section 2.5.4.5) in DAB are similar to traffic announcements (TAs) in RDS, although offering a wider range of categories such as news, weather information, events, etc. In total, 32 different categories have been defined. When designing a car receiver, it should be taken into account that TA is much more important for the user than, for example, event or finance information. Therefore, TA should be selectable preferably with only one button, whereas for other announcements it is tolerable, for the user to follow a menu structure to select these announcements.

8.8.3   Alternative Frequencies and Service Following

DAB provides frequency information about other DAB ensembles as well as FM or AM broadcast (see section 2.5.5). This serves as a tuning aid to the receiver control and is similar to the alternative frequency (AF) feature provided by RDS receivers. It offers the possibility to follow a service when the reception conditions of the actual frequency degrade too much. When the receiver leaves the coverage area of the DAB signal, it can automatically switch to the same programme being broadcast on an FM or AM frequency. It can also be used to switch (back) to DAB as soon as the receiver enters the coverage area. This feature is of particular interest in the transition phase from analogue to digital broadcasting.

Appendix 1

DAB Parameters for Modes I, II, III and IV

A1.1   System Parameters

Parameter	Mode I	Mode IV	Mode II	Mode III
Sub-carriers				
Number of sub-carriers: K	1536	768	384	192
Sub-carrier spacing: Δf	1 kHz	2 kHz	4 kHz	8 kHz
Time relations				
Transmission frame duration: T_{Frame}	96 ms	48 ms	24 ms	24 ms
	196608T *	98304T	49152T	49152T
Symbol duration: $T_{symOFDM} = T_{guard} + T_u$	1246 µs	623 µs	312 µs	156 µs
	2552T *	1276T	638T	319T
Guard interval duration: T_{guard}	246 µs	123 µs	62 µs	31 µs
	504T *	252T	126T	63T
Symbol duration without T_{guard}: $T_u = 1/\Delta f$	1000 µs	500 µs	250 µs	125 µs
	2048T *	1024T	512T	256T
Null-symbol duration: T_{null}	1297 µs	648 µs	324 µs	168 µs
	2656T *	1328T	664T	345T
OFDM symbols				
OFDM symbols per transmission frame (without null symbol): L	76	76	76	153
OFDM symbols with PR data	1	1	1	1
OFDM symbols with FIC data	3	3	3	8

Parameter	Mode I	Mode IV	Mode II	Mode III
OFDM symbols with MSC data	72	72	72	144

FIC/MSC

	Mode I	Mode IV	Mode II	Mode III
FIC: FIBs per transmission frame	12	6	3	4
FIBs per 24 ms frame	3	3	3	4
MSC: CIFs per transmission frame	4	2	1	1
CIFs per 24 ms frame	1	1	1	1
FIBs/CIF	3	3	3	4

Transmission frame

	Mode I	Mode IV	Mode II	Mode III
Bit per OFDM symbol	3.072 kbit	1.536 kbit	0.768 kbit	0.384 kbit
Bit per transmission frame (without PR symbol)	230.4 kbit	115.2 kbit	57.6 kbit	58.368 kbit
Transmission frames per second	10.416	20.832	41.666	41.666

Data rates

	Mode I	Mode IV	Mode II	Mode III
FIC data rate (gross, code rate always 1/3)	96 kbit/s	96 kbit/s	96 kbit/s	128 kbit/s
MSC data rate (gross)	2.304 Mbit/s	2.304 Mbit/s	2.304 Mbit/s	2.304 Mbit/s
Max. MSC net data rate for a single sub-channel **	1.824 Mbit/s	1.824 Mbit/s	1.824 Mbit/s	1.824 Mbit/s
Total data rate (with PR Symbol)	2.432 Mbit/s	2.432 Mbit/s	2.432 Mbit/s	2.448 Mbit/s

Network specific parameters

	Mode I	Mode IV	Mode II	Mode III
Maximum echo delay ($\approx 1.2 \times T_{guard}$)	300 μs	150 μs	75 μs	37.5 μs
Maximum propagation path difference	≈100 km	≈50 km	≈25 km	≈12.5 km
	(90 km)	(45 km)	(22.5 km)	(11.25 km)
Maximum f_{RF} ***	340 MHz		1.38 GHz	2.76 GHz
****	375 MHz		1.5 GHz	3.0 GHz

*   System clock: 2.048 MHz with a period T of 0.48828 μs.

**   The multiplex configuration for maximum data rate is as follows: one sub-channel with 1.824 Mbit/s and code rate 4/5 and a second sub-channel with 16 kbit/s and code rate 3/4. The remaining multiplex capacity of 64 bit per 24 ms frame is equivalent to an uncoded data rate of 8 kbit/s.

***   @ max f_{RF}: maximum S/N-degradation of 4 dB for a BER of 10^{-4} at a speed of 200 km/h and 1 dB at 100 km/h (Le Floch, 1992).

****   @ max f_{RF}: maximum S/N-degradation of 4 dB for a BER of 10^{-4} at a speed of 180 km/h and 1 dB at 90 km/h (Kozamernik, 1992).

A1.2   Important Relations:

1 CU	= 64 bits	= 8 bytes
1 CIF	= 864 CU = 55.296 kbits	= 6.912 kbytes
1 FIB	= 256 bits	= 32 bytes

Where: CU   = Capacity Unit

A1.3   Coarse Structure of the Transmission Frame:

|Null-symbol | PR | FIC (FIBs) | MSC (CIFs) |

Where: PR   = Phase Reference symbol
        FIC   = Fast Information Channel
        FIB   = Fast Information Block
        MSC   = Main Service Channel
        CIF   = Common Interleaved Frame

Appendix 2

DAB: Status of Introduction – World-wide

For more details see (www.WorldDab)

Region/Country	System	Frequency Band	Startup Date ***	Additional Information (number of channels, amount of coverage, etc.)
Europe				
Austria	DAB *	VHF/12D	(1999)	4 channels; ca. 19% of population
Belgium	DAB *		Pilot phase (2000)	7 channels; ca. 86% of population
Croatia	DAB *		(1997)	
Czech Republic	DAB *		Pilot phase	Ca. 12% of population
Denmark	DAB *		(1999)	Ca. 30% of population
Finland	DAB *	III	(1999)	10 channels; ca. 40% of population
France	DAB *		(1997)	Up to 13 channels in 2 multiplexes; ca. 25% of population
Germany	DAB *	L-band	(1997)	Ca. 100 channels in 16 different regions; ca. 30% of population
Hungary	DAB *		Pilot phase	4 channels; ca. 60% of population
Ireland	DAB *	VHF/12C	Pilot phase	6 services; ca. 30% of population
Israel	DAB *	VHF/12	Pilot phase	Ca. 85% of population
Italy	DAB *	VHF/12	(1995)	Ca. 33% of population
Norway	DAB *		(1999)	Ca. 35% of population (95% planned)
Poland	DAB *	II, III	Pilot phase	Ca. 8% of population
Portugal	DAB *	VHF/12B	Pilot phase	6 channels, 15 transmitters

Region/Country	System	Frequency Band	Startup Date ***	Additional Information (number of channels, amount of coverage, etc.)
Slovenia	DAB *		Test phase	
Spain	DAB *		(2000)	Ca. 30% of population, 2 multiplexes
Sweden	DAB *		(1995)	Ca. 85% of population, 3 networks
Switzerland	DAB *		(1999)	Ca. 55% of population, 2 multiplexes
The Netherlands	DAB *		Pilot phase	Ca. 45% of population
Turkey	DAB *	III	Planning phase	
United Kingdom	DAB *		(1995)	5 services, ca. 60% of population
Africa South Africa	DAB *		(1999)	7 services, ca. 10% of population
America Canada	DAB *		(1999)	Ca. 35% of population, including several PAD services
USA	other**			
Mexico	DAB *	L-band	Test phase	
Asia China	DAB *		(1996)	3 transmitters
Hong Kong	DAB *		Pilot phase	
India	DAB *	III	Test phase	(10 Million people, Delhi)
Japan	other**			
Malaysia	DAB *	III	Pilot phase	
Singapore	DAB *		(1999)	6 audio services, 10 multimedia services
South Korea	DAB *		Test phase	
Taiwan	DAB *	III	(2000)	Ca. 13% of population
Australia	DAB *	L-band	Pilot phase	5 networks; regular services planned to start 2001

*   Eureka 147 DAB system.
**   For details see section 1.5.2.
***   (year) = start of regular service.

Bibliography

Standards and Related Documents

(BO.789) ITU-R: Recommendation BO.789 (1994) Digital sound broadcasting to vehicular, portable and fixed receivers for broadcast satellite service (sound) in the frequency range 500 to 3000 MHz. Geneva.

(BO.1130) ITU-R: Recommendation BO.1130-2 (1998) Systems for digital sound broadcasting to vehicular, portable and fixed receivers for broadcasting – satellite service (sound) in the frequency range 1400 to 2700 MHz. Geneva.

(BS.645) ITU-R: Recommendation BS.645 (1994) Test signals and metering to be used on international sound-programme connections. Geneva.

(BS.774) ITU-R: Recommendation BS.774-2 (1997) Service requirements for digital sound broadcasting to vehicular, portable and fixed receivers in the VHF/UHF bands. Geneva.

(BS.775) ITU-R: Recommendation BS.775-1 (1997) Multichannel stereophonic sound system with and without accompanying picture. Geneva.

(BS.1114) ITU-R: Recommendation BS.1114-1 (1995) Systems for terrestrial digital sound broadcasting to vehicular, portable and fixed receivers in the frequency range 30 to 3000 MHz. Geneva.

(BS.1115) ITU-R: Recommendation BS.1115 (1995) Low bit-rate audio coding. Geneva.

(ITU-R, 1995) ITU-R: Special Publication (1995) Terrestrial and satellite digital sound broadcasting to vehicular, portable and fixed receivers in the VHF/UHF bands. Geneva.

(BPN 002) EBU document BPN 002 rev.1 (1995) DAB system: Definition of the Ensemble Transport Interface (ETI). *EBU Technical Texts*, Geneva.

(BPN 003) EBU document BPN 003 rev.1 (1998) Technical Basis for T-DAB Services Network Planning and Compatibility with existing Broadcasting Services; *EBU Technical Texts*, Geneva.

(BPN 007) EBU document BPN 007 (1996) A broadcasters introduction to the implementation of some key DAB system features – Part 1. *EBU Technical Texts*, Geneva.

(BPN 019) EBU document BPN19 (1999) Report on the EBU Subjective Listening Tests of Multichannel Audio Codecs. *EBU Technical Texts*, Geneva.

(BPN 022) EBU document BPN22 (1999) Practical Webcasting: The new range of opportunities for traditional broadcasters. *EBU Technical Texts*, Geneva.

(CEPT, 1995) CEPT (1995). The Final Acts of the CEPT T-DAB Planning Meeting, Wiesbaden.

(EN 50248) CENELEC: EN 50 248 (2001) Characteristics of DAB receivers. Brussels. *(also published as IEC 62 105)*

(EN 50255) CENELEC: EN 50 255 (1997) Digital Audio Broadcasting system – Specification of the Receiver Data Interface (RDI). Brussels. *(also published as IEC 62104)*

(EN 50067) ETSI: EN 50 067 (1992) Specification of the Radio Data System (RDS) for VHF/FM sound broadcasting in the frequency range from 87,5 to 108,0 MHz. Sophia-Antipolis.

(EN 50320) CENELEC: EN 50 320 (2000) DCSR (DAB Command Set for Receiver). Brussels.

(ETS 300401) former ETSI standard, now replaced by EN 300 401, see below.

(EN 300401) ETSI: EN 300 401 3rd edition (2000) Radio broadcasting systems: Digital Audio Broadcasting (DAB) to mobile, portable and fixed receivers. Sophia-Antipolis.

(EN 300421)   ETSI: EN 300 421 (1994) Digital broadcasting systems for television, sound and data services; Framing structure, channel coding and modulation for 11/12 GHz satellite systems. Sophia-Antipolis.

(EN 300429)   ETSI: EN 300 429 (1994) Digital broadcasting systems for television, sound and data services; Framing structure, channel coding and modulation for cable systems. Sophia-Antipolis.

(EN 300744)   ETSI: EN 300 744 (1996) Digital broadcasting systems for television, sound and data services; Framing structure, channel coding and modulation for digital terrestrial television. Sophia-Antipolis.

(EN 300797)   ETSI: EN 300 797 V1.1.1 (1999) Digital Audio Broadcasting (DAB); Distribution interfaces; Service Transport Interface (STI). Sophia-Antipolis.

(EN 300798)   ETSI: EN 300 798 V1.1.1 (1998) Digital Audio Broadcasting (DAB); Distribution interfaces; Digital base-band In-phase and Quadrature (DIQ) Interface. Sophia-Antipolis.

(EN 300799)   ETSI: ETS 300 799 1st edition (1997) Digital Audio Broadcasting (DAB); Distribution interfaces; Ensemble Transport Interface (ETI). Sophia-Antipolis.

(EN 301234)   ETSI: EN 301 234 V1.2.1 (1999) Digital Audio Broadcasting (DAB); Multimedia Transfer Protocol (MOT). Sophia-Antipolis.

(EN 301700)   ETSI: EN 301 700 (2000) Digital Audio Broadcasting (DAB); VHF/FM Broadcasting: cross-referencing to simulcast DAB services by RDS-ODA 147. Sophia-Antipolis.

(ES 201735)   ETSI: ES 201 735 (TS 101 735) (2000) Digital Audio Broadcasting (DAB); IP Datagram Tunnelling. Sophia-Antipolis.

(ES 101736)   ETSI: ES 101 736 (2000) Digital Audio Broadcasting (DAB); Network Independent Protocols for Interactive Services. Sophia-Antipolis.

(ES 201737)   ETSI: ES 201 737 (TS 201 737) (2000) Digital Audio Broadcasting (DAB); Interaction Channel through GSM/PSTN/ISDN/DECT. Sophia-Antipolis.

(ES 201755)   ETSI: ES 201 755 (TS 101 500) (2000) Digital Audio Broadcasting (DAB); Multichannel Audio. Sophia-Antipolis. *(also published as TS 101 500)*

(H.263)   ITU-T: Recommendation H.263 (1998). Video coding for low bit-rate communication. Geneva.

(IEC 60268)   IEC Publ. 60268-10 (1991) Sound system equipment – Part 10: Peak programme level meters; IEC Publ. 60268-17 (1990) Sound system equipment – Part 17: Standard volume indicators; IEC Publ. 60268-18 (1995) Sound system equipment – Part 18: Peak programme level meters - Digital audio peak level meters. Geneva.

(IEC 60958)   IEC 60958-1 (1999) Digital audio interface – Part 1: General; IEC 60958-2 Part 2: Software information delivery mode; IEC 60958-3 (1999) Part 3: Consumer applications; IEC 60958-4 (1999) Part 4: Professional applications.

(IEC 61937)   IEC 61937 (2000) Digital audio – Interface for non-linear PCM encoded audio bit-streams applying IEC 60958. Geneva.

(IS 11172)   ISO/IEC: International Standard IS 11172 (1993) Information technology – Coding of moving pictures and associated audio for digital storage media up to about 1.5 Mbit/s (*MPEG-1*). Part 1: Video; Part 3: Audio. Geneva.

(IS 13818)   ISO/IEC: International Standard IS 13818-3 (1998) Information technology – Coding of moving pictures and associated audio information – Part 3: Audio (*MPEG-2*). Geneva.

(J.52)   ITU-T: Recommendation J.52 (1992) Digital Transmission of High-Quality Sound-Programme Signals using one, two or three 64 kbit/s Channels per Mono Signal (and up to six per Stereo Signal). Geneva.

(P.370)   ITU-R: Recommendation P.370-7 (1998) VHF and UHF Propagation Curves for the Frequency Range from 300 MHz to 1000 MHz. Geneva.

(R68-1992)   EBU: Technical Recommendation R68-1992 (1992) Alignment level in digital
                audio production equipment and in digital audio recorders. *EBU Techn. Texts,*
                Geneva.

(R96-1999)   EBU: Technical Recommendation R96-1999 (1999) Formats for production
                and delivery of multichannel audio programmes. *EBU Techn. Texts,* Geneva.

(TR-004)   EACEM: TR-004 (2000) Application of the EMC Directive 89/336/EEC for
                Digital Audio Broadcast receivers.

(TR 101495)   ETSI: TR 101 495 (2000) DAB; Guide to standards, guidelines and
                bibliography. Sophia-Antipolis. (*submitted to ETSI)*

(TR 101496)   ETSI: TR 101 496 (2000) Digital Audio Broadcasting (DAB): Guidelines and
                rules for implementation and operation. Sophia-Antipolis.

(TR 101497)   ETSI: TR 101 497 (xxxx) Digital Audio Broadcasting (DAB): Rules of
                operation for the Multimedia Object Transfer Protocol. Sophia-Antipolis. *(still
                under consideration)*

(TR 101758)   ETSI: TR 101 758 (2000) Digital Audio Broadcasting (DAB): DAB signal
                strength and receiver parameter; Targets for typical operation. S.-Antipolis.

(TS 101498)   ETSI: TS 101 498 (2000) Digital Audio Broadcasting (DAB): Broadcast Web
                Site Application. Sophia-Antipolis.

(TS 101499)   ETSI: TS 101 499 (2000) Digital Audio Broadcasting (DAB): Slide Show
                Application. Sophia-Antipolis.

(TS 101500)   ETSI: TS 101 500 (2000) Digital Audio Broadcasting (DAB); Multichannel
                Audio. Sophia-Antipolis. *(see also (ES 201755))*

(TS 101735)   ETSI: TS 101 735 (2000) Digital Audio Broadcasting (DAB); Internet
                Protocol (IP) Datagram Tunnelling. Sophia-Antipolis.

(TS 101736)   ETSI: TS 101 736 (2000) Digital Audio Broadcasting (DAB); Network
                Independent Protocols for Interactive Services. Sophia-Antipolis.

(TS 101737)   ETSI: TS 101 737 (2000) Digital Audio Broadcasting (DAB); DAB
                Interaction Channel through GSM/PSTN/ISDN/DECT. Sophia-Antipolis.

(TS 101756)   ETSI: TS 101 756 (2000) Digital Audio Broadcasting (DAB); Registered
                Tables. Sophia-Antipolis.

(TS 101757)   ETSI: TS 101 757 (xxxx) Digital Audio Broadcasting (DAB); Conformance
                Testing for DAB Audio. Sophia-Antipolis. *(still under consideration)*

(TS 101759)   ETSI: TS 101 759 (xxxx) Digital Audio Broadcasting (DAB); Transparent
                Data Channel. Sophia-Antipolis. *(still under consideration)*

(TS 101860)   ETSI: TS 101860 (xxxx) Digital Audio Broadcasting (DAB); Distribution
                Interfaces; Service Transport Interface (STI); STI Levels. Sophia-Antipolis.
                (still under consideration)

(UECP)   EBU RDS Forum: SPB 490 (1997) RDS Universal Encoder Communication
                Protocol (UECP). Geneva.

Key

BO.   ITU-R Recommendation (Satellite sound and television broadcasting)
BPN   EBU: Internal Technical Document (informative)
BR.   ITU-R Recommendation (Sound and television recording)
BS.   ITU-R: Recommendation (Sound broadcasting)
EN   ETSI or CENELEC European Telecommunication Standard (normative)
ES   ETSI: Standard (normative)
IS   ISO/IEC: International Standard (normative)
P.   ITU-R Recommendation (Radiowave propagation)
R   EBU: Technical Recommendation
TR   ETSI: Technical Report (informative)
TS   ETSI: Technical Specification (normative)

Publications

(AES, 1996)   AES Special publication: Gilchrist, N. H. G. and Grewin, C. (editors) (1996)
                     Collected papers on digital audio bit-rate reduction, AES, New York.
(ARD, 1994)   ARD-Pflichtenheft 5/3.8 Teil 1 (1994) RDS Transmission Protocol: Coder für
                     Radiodatensystem (RDS) – Zusatzinformationen für den Hörrundfunk gemäß
                     DIN EN 50 067. Institut für Rundfunktechnik, München.
(Atkinson, 1998)   Atkinson, S. and Strange, J. (1998) Integration Trends with RF Silicon
                     Technologies in Mobile Radio Applications. *Proc. IEEE RFIC Symposium*,
                     Baltimore, MD, 83–86
(Bolle, 1997)   Bolle, M. (1997) Verfahren zur automatischen Frequenzkorrektur in
                     Mehrträgerübertragungsverfahren, Pat. DE 4 441 566
(Bolle, 1998)   Bolle, M., et al. (1998) Architecture and Performance of an Alternative DAB
                     Receiver Chipset. *IEEE VLSI Conference*, Honolulu.
(Bosch, 1999)   Bosch (1999) Research Info, No. IV/1999, Hildesheim.
(Bossert, 1998)   Bossert, M. (1998) *Kanalcodierung*, 2nd edition, Teubner-Verlag, Stuttgart.
(Chambers, 1992)   Chambers, J.P. (1992) DAB system multiplex organisation. *Proc. 1st EBU
                     International Symposium on Digital Audio Broadcasting*, Montreux, 111–120
(Clark, 1988)   Clark, G.C. and Cain, J.B. (1988) *Error-Correction Coding for Digital
                     Communications*, Plenum, New York.
(Clawin,1998)   Clawin, D., et al. (1998) Architecture and Performance of an Alternative DAB
                     Receiver Chip Set. *Proc. 28th European Microwave Conference,* Amsterdam
                     1998, vol.2, 645–650
(Clawin,2000)   (Clawin,, D., Gieske, K., Hofmann, F., Kupferschmidt, C., Mlasko, T.,
                     Naberfeld, F., Passoke, J., Spreitz, G. and Stepen, M. (2000) D-FIRE 2: A
                     combined DAB/FM receiver system on chip. *MicroTec 2000*, Hannover.
(Culver, 1996)   Culver, R.D. (1996). Report of the Field Test Task Group. *Consumer
                     Electronics Manufacturers Association*, Washington, DC
(David, 1996)   David, K. and Benkner, Th. (1996) *Digitale Mobilfunksysteme*, Teubner-
                     Verlag, Stuttgart.
(Dehéry, 1991)   Dehéry, Y.F., Stoll, G. and van de Kerkhof, L. (1991) MUSICAM Source
                     Coding for Digital Sound. *Symposium Record "Broadcast Sessions" of the
                     17th International Television Symposium*. Montreux, 612–617
(Dietz, 2000)   Dietz, M. and Mlasko, T. (2000) Using MPEG-4 audio for DRM digital
                     narrowband broadcasting. *Proc. IEEE Circuits & Systems Conference 2000*
(EBU, 1999)   EBU document: B/CASE 100/BMC 477 (1999) Subjective Audio Quality
                     Achievable at Various Bit-rates for MPEG-Audio Layer II and Layer III.
                     European Broadcasting Union, Geneva.
(Fastl, 1977)   Fastl, H. (1977) Temporal Masking Effects: II. Critical Band Noise Masker.
                     Acustica **36**, 317
(Fettweis, 1986)   Fettweis, A. (1986) Wave Digital Filters – Theory and Practice. *Proc. IEEE*,
                     vol. 25, no. 2, 270–326
(Frieling, 1996)   Frieling, G., Lutter, F.J. and Schulze, H. (1996) Bit error measurements for
                     the assessment of a DAB radio supply. *Rundfunktech. Mitteilungen* **40**, 4, 121
(Gilb,2000)   Gilb, J. (2000) Bluetooth Radio Architectures. *RFIC Symposium 2000, Digest
                     of Papers*, Boston, USA, 3–6
(Gilchrist, 1998)   Gilchrist, N.H.G. (1998) Audio levels in digital broadcasting. *105th AES
                     Convention*, San Francisco, Preprint 4828
(Goldfarb, 1998)   Goldfarb, M., Croughwell, R., Schiller, C., Livezey, D. and Heiter G. (1998)
                     A Si BJT IF Downconverter AGC IC for DAB. *Digest of 1998 IEEE MTT-S
                     International Microwave Symposium*, Baltimore, vol.1, 353–356
(Hagenauer, 1988)   Hagenauer, J. (1988) Rate compatible punctured convolutional RCPC codes
                     and their applications, *IEEE Trans. Commun. COM-36,* 389–400

(Hallier, 1994a)   Hallier, J. et al. (1994) Preparing Low-Cost ICs for DAB: The JESSI-DAB
                        Programme. *Proc. 2nd International Symposium. on DAB.* Toronto.

(Hallier, 1994b)   Hallier, J., Lauterbach, Th. and Unbehaun, M. (1994) Multimedia
                        broadcasting to mobile, portable and fixed receivers using the Eureka 147
                        Digital Audio Broadcasting System. *Proc. ICCC Regional Meeting on
                        Wireless Computer Networks (WCN),* Den Haag, 794

(Hallier, 1995)   Hallier, J., Lauterbach, Th. and Unbehaun, M. (1995) Bild- und
                        Videoübertragung über DAB – Ein erster Schritt zum Multimedia-Rundfunk,
                        ITG-Fachbericht 133 Hörrundfunk, VDE-Verlag, Berlin, 149

(Hoeg, 1994)   Hoeg, W., Gilchrist, N., Twietmeyer, H. and Jünger, H. (1994) Dynamic
                        Range Control (DRC) and Music/Speech Control (MSC) as additional data
                        services for DAB. *EBU Techn. Rev.* **261**, 56

(Hoeher, 1991a)   Hoeher, P. (1991) TCM on frequency-selective land-mobile fading channels.
                        Proc. 5th Tirrenia Int. Workshop on Digital Communication.

(Hoeher, 1991b)   Hoeher, P., Hagenauer, J., Offer, E., Rapp, C. and Schulze, H. (1991)
                        Performance of an RCPC-coded OFDM-based Digital Audio Broadcasting
                        (DAB) System, *Proc. IEEE Globecom 1991 Conference* vol.1, 2.1.1–2.1.7

(Hofmeier, 1995)   Hofmeier, St. (1995) Der Countdown für das digitale Astra-Radio läuft.
                        Funkschau **2**, 42

(Ilmonen, 1971)   Ilmonen, K. (1971) Listener preferences for loudness balance of broadcasts,
                        with special consideration to listening in noise. *Research Report from Finnish
                        Broadcasting Company,* Helsinki, no. 12/1971

(Imai, 1977)   Imai, H.; and Hirakawa, S. (1977) A new multilevel coding method using
                        error correcting codes. *IEEE Trans. Inf. Theory IT-23,* 371–377

(IRT, 1998)   Institut für Rundfunktechnik (1998) Empfehlung 14 KSZH
                        Funkhaustelegramme – Aufbau und Anwendung. München.

(IRT, 1999)   Institut für Rundfunktechnik (1999) Multimedia Object Transfer Protocol –
                        Interface. München.

(ITTS, 1994)   Interactive Text Transmission System, EACEM Tech. Doc. no. 7

Jongepier, 1996)   Jongepier, B. (1996) The JESSI DAB Chipset. *Proc. AES UK DAB
                        Conference,* London, 76

(Kammeyer, 1996)   Kammeyer, K.-D. (1996) *Nachrichtenübertragung,* 2nd edition, Teubner-
                        Verlag. Stuttgart.

(Kate, 1992)   ten Kate, W., Boers, P., Mäkivirta, A., Kuusama, J., Christensen, K.-E. and
                        Sørensen, E. (1992) Matrixing of Bit rate Reduced Signals. *Proc. ICASSP,*
                        vol. 2, 205–208

(Kleine, 1995)   Kleine, G. (1995) Astra Digital Radio. *Funkschau* **10**, 44

(Klingenberg, 1998)   Klingenberg, W. and Neutel, A. (1998) MEMO: A Hybrid DAB/GSM
                        Communication System for Mobile Interactive Multimedia Services. In:
                        Multimedia Applications, Services and Techniques – Proc. ECMAST '98,
                        Springer-Verlag, Berlin 1998 (Lecture Notes in Computer Science, vol. 1425)

(Kopitz, 1999)   Kopitz, D. and Marks, B. (1999) Traffic and travel information broadcasting –
                        protocols for the 21st century, *EBU Techn. Rev.* **279**, 1

(Korte, 1999)   Korte, O. and Prosch, M. (1999) Packet Mode Inserter: Konzepte zur
                        Realisierung eines Multimedia Datenservers für den digitalen Rundfunk.
                        Fraunhofer Institut für Integrierte Schaltungen, Erlangen.

(Kozamernik, 1992)   Kozamernik, F. (1992) Digital Sound Broadcasting Concepts. *Proc. Europa
                        Telecom '92,* Budapest.

(Kozamernik, 1999)   Kozamernik, F. (1999). Digital Audio Broadcasting – coming out of the
                        tunnel. *EBU Techn. Rev.* **279**

(Kuroda, 1997)   Kuroda, T. et al. (1997) Terrestrial ISDB System using Band Segmented
                        Transmission Scheme. *NHK Laboratories note* no. 448

(Lauterbach, 1996)   Lauterbach, Th., Stierand, I., Unbehaun, M. and Westendorf, A. (1996) Realisation of Mobile Multimedia Services in Digital Audio Broadcasting (DAB), *Proc. European Conference on Multimedia Applications, Services and Techniques,* Part II, 549

(Lauterbach, 1997a)   Lauterbach, Th., Unbehaun, M., Angebaud, D., Bache, A., Groult, Th., Knott, R., Luff, P., Lebourhis, G., Bourdeau, M., Karlsson, P., Rebhan, R., Sundström, N. (1997) Using DAB and GSM to provide Interactive Multimedia Services to Portable and Mobile Terminals. In: Serge Fdida and Michele Morganti (eds) *Multimedia applications, Services and Techniques, Proc. ECMAST,* Milan, Springer-Verlag, Berlin.

(Lauterbach, 1997b)   Lauterbach, Th. and Unbehaun, M. (1997) Multimedia Environment for Mobiles (MEMO) – Interactive Multimedia Services to Portable and Mobile Terminals, *ACTS Mobile Communications Summit,* Aalborg, vol. II, 581

(Layer, 1998)   Layer, F., Englert, T., Friese, M. and Ruf, M. (1998) Locating Mobile Receivers using DAB Single Frequency Networks. *Proc. 3rd ACTS Mobile Communication Summit,* Rhodos, 592–597

(Le Floch, 1992)   Le Floch, B. (1992) Channel Coding and Modulation for DAB. *Proc. 1st International Symposium on DAB,* Montreux.

(Lever, 1997)   Lever, M., Richard, J. and Gilchrist N.H.C. (1997) Subjective assessment of the error performance of the Eureka 147 DAB system. *102nd AES Convention,* Munich, Preprint 4483

(Liu, 2000)   Liu, E. et al. (2000) A 3 V Mixed-Signal Baseband Processor IC for IS-95. *ISSCC 2000, Digest of Technical Papers*

(Marks, 2000)   Marks, B. (2000) TPEG applications, compatible with RDS-TMC, give greater richness and wider content. *2nd WorldDAB Multimedia Conference,* London (available electronically at (www.worlddab) only)

(Meares, 1998)   Meares, D.J. and Theile,G. (1998) Matrixed Surround Sound in an MPEG Digital World. *J. AES* **46**, 4, 331–335

(Müller, 1970)   Müller, K. (1970) What dynamic range does the listener wish to have? *EBU Rev.* **124-A**, 269

(Nakahara, 1996)   Nakahara, S. et al. (1996) Efficient Use of Frequencies in Terrestrial ISDB System. *IEEE Trans. Broadcast.* **42**, no. 3, 173

(Nowottne, 1998)   Nowottne, H.-J., Heinemann, C., Peters, W. and Tümpfel, L. (1998) DAB-Programmzuführung – Neue Anforderungen und ihre Realisierung. *Proc. 13th International Scientific Conference,* Mittweida, Germany, **D**, 41–48

(Paiement, 1996)   Paiement, R. (1996) Description and early measurement results: experimental distributed emission DSB network at L-Band. *Document submitted to ITU-R WP 10B (doc. 10B/16),* Toledo.

(Paiement, 1997)   Paiement, R. (1997) Update and further results, experimental DSB system at L-Band. *Document submitted to ITU-R WP 10B (doc. 10B/41).* Geneva

(Par, 1994)   Van de Par, S. (1994) On 2 1/2-channel sound compared to three-channel sound. *J. AES* **42**, 555–564

(Peters, 1999)   Peters, W., Nowottne, H.-J., Madinger, R., Sokol, P.and Tümpfel, L. (1999) DAB-Programmzuführung im Regelbetrieb – Realisierung mit dem neuen Standard STI. *ITG Fachbericht 158, Fachtagung Hörrundfunk,* Cologne, 145

(Plenge, 1991)   Plenge, G. (1991) DAB – Ein neues Hörrundfunksystem. *Rundfunktech. Mitteilungen,* 45

(Proakis, 1995)   Proakis, J.G. (1995) *Digital Communications.* 3rd edition, McGraw-Hill, NewYork.

(Reimers, 1995)   Reimers, U. (1995) *Digitale Fernsehtechnik,* Springer-Verlag, Berlin

(Riley, 1994)   Riley, J.L. (1994). The DAB Multiplex and system support features. *EBU Tech. Rev.* **259**, 11

(Sachdev, 1997)   Sachdev, D.K. (1997) The WorldSpace System: Architecture, Plans and Technologies. *Proc. 51st NAB Conference,* Las Vegas, 131

(Schambeck, 1987)   Schambeck, W. (1987) Digitaler Hörfunk über Satelliten. In: Pauli, P.(ed.)
                      Fernseh- und Rundfunk-Satellitenempfang, Expert-Verlag, Ehningen b.
                      Böblingen, S. 145–177

(Schramm, 1997)   Schramm, R. (1997) Pseudo channel bit error rate as an objective factor for
                      the assessment of the supply with DAB signals. *EBU Tech. Rev.* **274**

(Schulze, 1995)   Schulze, H. (1995) Codierung, Interleaving und Multiplex-Diversity bei DAB:
                      Auswirkungen auf die Rundfunkversorgung, *ITG Fachbericht 135 Mobile
                      Kommunikation*, 477-484

(Schulze, 1998)   Schulze, H. (1998) Möglichkeiten und Grenzen des Mobilempfangs von
                      DVB-T, *3. OFDM-Fachgespräch*, Braunschweig

(Schulze, 1999)   Schulze, H. (1999) The Performance of a Coded Multicarrier 64-QAM
                      System with Channel Estimation, *Proc. 1st International OFDM-Workshop*,
                      Hamburg, 15.1–15.4

(Sicre, 1994)   Sicre J.L., et al. (1994) Preparing Low-Cost ICs for DAB: The JESSI-DAB
                      Programme, *Proc. 2nd International Symposium on DAB*, Toronto

(Smith, 1989)   Smith, A. B. and Snyder, C. D. (1989) How the audio quality depends on the
                      bit rate. *J. AES* **45**, 123–134

(Stoll, 1995)   Stoll, G. (1995) Audio compression schemes used in DAB: The MPEG-1 and
                      MPEG-2 Audio Layer II coding standard for two- and multi-channel sound.
                      Proc. EBU/EuroDab In-depth Seminar on DAB, Montreux.

(Stoll, 2000)   Stoll, G. and Kozamernik, F. (2000) EBU Listening tests on Internet audio
                      codecs. *EBU Techn. Rev.* **283**

(Stott, 1996)   Stott, J.H. (1996) The DVB terrestrial (DVB-T) specification and its
                      implementation in a practical modem. *International Broadcasting Convention,
                      Conference Publication*, no. 428

(Strange, 2000)   Strange, J. and Atkinson, S. (2000) A Direct Conversion Transceiver for
                      Multi-Band GSM *Application, RFIC Symposium 2000, Digest of Technical
                      Papers*, Boston, 25–28

(STRL, 1999)   NHK Science and Technical Research Laboratories (1999) *STRL Newsletters*,
                      nos 3 and 5

(Taddiken, 2000)   Taddiken, et al. (2000) Broadband Tuner on a Chip for Cable Modem, HDTV,
                      and Legacy Analog Standards. *RFIC Symposium 2000, Digest of Technical
                      Papers,* Boston.

(TEMIC, 1999)   U2731B Data-Sheet (1999) TEMIC-Semiconductor GmbH. Heilbronn

(Theile, 1993)   Theile, G. and Link, M. (1993) Low-complexity Dynamic Range Control
                      system based on scale-factor weighting. *94th Convention AES*, Berlin,
                      Preprint 3563

(Titus, 1998)   Titus, W., Croughwell, R., Schiller, C. and DeVito, L. (1998) A Si BJT Dual
                      Band Receiver IC for DAB. *Digest of 1998 IEEE MTT-S International
                      Microwave Symposium*, Baltimore, USA, vol. 1, 345–348

(U2731B)   U2731B Data-Sheet, TEMIC-Semiconductor GmbH, Heilbronn.

(Voyer, 1996)   Voyer, R. and Breton, B. (1996) Coverage Prediction for distributed emission
                      of DAB. *Proc.Third International Symposium on Digital Audio Broadcasting*,
                      Montreux.

(Waal, 1991)   van der Waal, R.G. and Veldhuis, R.N.J. (1991) Sub-band Coding of
                      Stereophonic Digital Audio Signals, *Proc. ICASSP*

(Watkinson, 2000)   Watkinson, (2000) *The Art of Digital Audio*, 3rd edition, Butterworth-
                      Heinemann, Oxford.

(Weck, 1995)   Weck, Ch. (1995) The Error Protection of DAB. *Records of AES UK
                      Conference "DAB, The Future of Radio"*, London, 23–32

(Wiese, 1992)   Wiese, D. (1992) Optimisation of Error Detection and Concealment for
                      Strategies for Digital Audio Broadcasting (DAB). *92nd Convention AES*,
                      Vienna, Preprint 3264

(Wörz, 1993)   Wörz, Th. and Hagenauer, J. (1993) Decoding of M-PSK-multilevel-Codes.
                        ETT **4**, 299–308
(Wüstenhagen, 1998)   Wüstenhagen, U., Feiten, B. and Hoeg, W. (1998) Subjective Listening Test
                        of Multichannel Audio Codecs. *105th Convention AES, San Francisco,*
                        Preprint 4813
(Zwicker, 1967)   Zwicker, E. and Feldtkeller, R. (1967) *Das Ohr als Nachrichtenempfänger.*
                        2nd edition, Hirzel-Verlag, Stuttgart.

Further Reading

AES Special publication (1996) Gilchrist, N.H.G. and Grewin, C. (eds) *Collected papers on digital audio bit-rate reduction*, AES, New York.

Collins, G.W. (2000) *Fundamentals of Digital Television Transmission.* John Wiley & Sons, Chichester.

De Gaudenzi, R. and Luise, M. (eds) (1994) *Audio and Video Digital Broadcasting Systems and Techniques.* Elsevier, Amsterdam.

Eureka 147 Project (1997) *Digital Audio Broadcasting.* Brochure prepared for WorldDAB, Geneva

ITU-R Special Publication (1995) *Terrestrial and satellite digital sound broadcasting to vehicular, portable and fixed receivers in the VHF/UHF bands.* Radiocommunication Bureau, Geneva.

Lauterbach, T. (ed.) (1996) *Digital Audio Broadcasting: Grundlagen, Anwendungen und Einführung von DAB*, Franzis-Verlag, München.

Menneer, P. (1996) The Market for DAB – Sources and Findings of existing Market Research Studies, Prepared for EuroDAB Forum, Geneva.

Müller, A., Schenk, M. and Fugmann, J. (1998) *Datendienste im Digital Audio Broadcasting DAB,* Schriftenreihe der LFK – Band 2, Neckar-Verlag, Ulm.

Müller-Römer, F. (1998) *Drahtlose terrestrische Übertragung an mobile Empfänger.* Schriftenreihe der SLM, Band 4, Sächsische Landesanstalt für privaten Rundfunk und neue Medien (SLM), VISTAS-Verlag, Berlin.

Tvede, L., Pircher, P. and Bodenkamp, J. (1999) *Data Broadcasting – The Technology and the Business.* John Wiley & Sons, Chichester.

Internet Links

Note: The reader should be aware that any reference to an Internet web-page will be of informal character only. The content of the cited web-page may be subject to change.

(www.ASTRA)   Web-site of SES-ASTRA Digital Satellite Radio.
                  URL: http://www.ses-astra.com
(www.Siriusradio) Web-site of Sirius Satellite Radio, Inc.
                  URL: http://www.siriusradio.com
(www.WorldDAB)   Web-site of the WorldDAB Organisation, Geneva, Switzerland.
                  URL: http://www.worlddab.org
(www.XMRadio)   Web-site of XM Satellite Radio.
                  URL: http://www.XMradio.com
(www.mobil-info) Web-site of Mobil-Info service.
                  URL: http://www.mobil-info.de
(www.m4mforum)   Web-site of the M4M (multimedia for mobiles) Forum
                  URL: http://www.m4mforum.org

Index

DATE DUE

GAYLORD

PRINTED IN U.S.A.